Alone in the Universe

JOHN GRIBBIN

Alone in the Universe
Why Our Planet Is Unique

John Wiley & Sons, Inc.

Copyright © 2011 by John and Mary Gribbin. All rights reserved

Published by John Wiley & Sons, Inc., Hoboken, New Jersey

First published in Great Britain in 2011 by Penguin Books Ltd.

No part of this publication may be reproduced, stored in a retrieval system, or transmitted in any form or by any means, electronic, mechanical, photocopying, recording, scanning, or otherwise, except as permitted under Section 107 or 108 of the 1976 United States Copyright Act, without either the prior written permission of the Publisher, or authorization through payment of the appropriate per-copy fee to the Copyright Clearance Center, 222 Rosewood Drive, Danvers, MA 01923, (978) 750-8400, fax (978) 646-8600, or on the web at www.copyright.com. Requests to the Publisher for permission should be addressed to the Permissions Department, John Wiley & Sons, Inc., 111 River Street, Hoboken, NJ 07030, (201) 748-6011, fax (201) 748-6008, or online at http://www.wiley.com/go/permissions.

Limit of Liability/Disclaimer of Warranty: While the publisher and the author have used their best efforts in preparing this book, they make no representations or warranties with respect to the accuracy or completeness of the contents of this book and specifically disclaim any implied warranties of merchantability or fitness for a particular purpose. No warranty may be created or extended by sales representatives or written sales materials. The advice and strategies contained herein may not be suitable for your situation. You should consult with a professional where appropriate. Neither the publisher nor the author shall be liable for any loss of profit or any other commercial damages, including but not limited to special, incidental, consequential, or other damages.

For general information about our other products and services, please contact our Customer Care Department within the United States at (800) 762-2974, outside the United States at (317) 572-3993 or fax (317) 572-4002.

Wiley also publishes its books in a variety of electronic formats and by print-on-demand. Some content that appears in standard print versions of this book may not be available in other formats. For more information about Wiley products, visit us at www.wiley.com.

ISBN 978-1-118-14797-9 (cloth); ISBN 978-1-118-17539-2 (ebk);
ISBN 978-1-118-17540-8 (ebk); ISBN 978-1-118-17541-5 (ebk)

Printed in the United States of America

10 9 8 7 6 5 4 3 2 1

It seems almost as if our Galaxy were a giant warehouse containing the spare parts needed for life.

James Lovelock, FRS

For Simon Goodwin,
who was generous enough not to write it first!

Contents

Acknowledgements   xi
Preface: The Only Intelligent Planet   xiii

Introduction: One in a Trillion   1
   Across the Milky Way; Hot jupiters; Planets in profusion; Dusty beginnings; Cosmic chemistry; The life of Gaia; Searching for other Gaias

1. Two Paradoxes and an Equation   26
   The cosmic lottery and the Drake equation; The inspection paradox and the Copernican principle; Panspermia and the Fermi paradox; Probing for an answer

2. What's So Special about Our Place in the Milky Way?   55
   Making galaxies; Making metals; Mixing metals in the Milky Way; Our place in the Milky Way; The Galactic Habitable Zone; Catastrophic comets

3. What's So Special about the Sun?   80
   The narrow zone of life; The Sun is not an average star; Perturbing partners; Blasts from the past; The mystery of solar metallicity; Until the Sun dies; Postponing Doomsday

4. What's So Special about the Solar System?   100
   Too hot to handle; The geography of the Solar System; Making planets; Making the Solar System; Making the Earth; The special one

5. What's So Special about the Earth?   126
   Like a diamond in the sky; A planetary jigsaw puzzle; Creating continents; A field of force; Venus and Mars; A planetary stabilizer; Plate tectonics and life

CONTENTS

6. What's So Special about the Cambrian Explosion?   151
   I. Contingency and Convergence
   The Cambrian explosion; The Burgess Shale; Contingency;
   Convergence; The third way

7. What's So Special about the Cambrian Explosion?   167
   II. Hothouse Venus/Snowball Earth
   After the deep freeze; Tipping the balance; From without or within?;
   The archetypal impact; Cosmic clouds and comet dust; Diamond
   dust and a facelift for a goddess

8. What's So Special about Us?   184
   Chance, necessity and the decimal system; The molecular clock; The
   trigger for change; The pacemaker of human evolution; The fate of
   technological civilization; The fate of the Earth; No second chance

Further Reading   206
Index   211

Acknowledgements

Thanks to Simon Goodwin, Douglas Lin, Charley Lineweaver, Jim Lovelock, Mac Low, Josep Trigo-Rodriguez and the Sussex University astronomers for discussions and advice on various aspects of this story. Sharing an office with Bernard Pagel in recent years gave me an opportunity to learn more than I realized at the time about the chemical composition of the stars and how it changes as time passes. As ever, Mary Gribbin played a key role in ensuring that my words are intelligible, and the Alfred C. Munger Foundation provided a contribution to our travel and other expenses.

Preface
The Only Intelligent Planet

Do we owe our existence to the impact of a 'supercomet' with Venus 600 million years ago? A decade ago the idea might have seemed laughable. But now we know that there are icy objects the size of Pluto in orbits far out on the fringe of the Solar System, we know that Earth's Moon was formed by the impact of an object the size of Mars with Earth, we know from chaos theory that no orbit around the Sun is stable, and, crucially, we know that something catastrophic happened to both Venus and Earth at the same time, just before the explosion of life on Earth that led to our existence. Coincidence? Maybe. But if so, it is the most significant in a chain of coincidences that led to the emergence of intelligent life on Earth. And that chain has so many weak links that it may mean that, for all the proliferation of stars and planets in the Universe, as an intelligent species we may be unique.

There are several hundred billion stars in our Milky Way Galaxy. At a conservative estimate, several billion of them are orbited by planets capable of supporting life. There may be more habitable planets in the Galaxy than there are people on planet Earth. But 'habitable' does not mean 'inhabited'. The thesis of this book is that intelligent civilization exists only on Earth. The reason is connected with the series of cosmic events that affected both Venus and Earth some 600 million years ago. But this is only part of the story, just one of the astronomical and geophysical reasons why the Earth is special, and probably unique.

Since the time of Copernicus, the progress of science has resulted in a steady displacement of the perceived place of human beings from centre stage in the Universe. By the end of the twentieth century, the received wisdom was that we are an ordinary kind of animal living on an ordinary planet orbiting an ordinary star in the backwoods of an unspectacular galaxy. But this image of the Earth and humankind as an insignificant unit in the Universe may be wrong. This book challenges that idea and suggests that human beings are special after all – the unique products of an extraordinary set of circumstances that have as yet occurred nowhere else in the Galaxy, and possibly not in the entire Universe. An idea known as the 'Goldilocks Effect' says that there is something odd about our entire Universe; but such speculations have no place in my argument; whether or not the Universe is unusual, there is something odd about the place of the living Earth within the Universe.

The idea of Earth as a 'living planet', expressed most clearly by the concept of Gaia, has captured the imagination of a wide public and become respectable science. We are all used to the idea that our home in space is one interlocking system of life, and have become alerted to the very real prospect that human activities may be the death of Gaia. Such ideas are discussed in recent books from Jim Lovelock. The picture he paints is bad enough from the parochial perspective of life on Earth itself. But does one planet really matter among the immensity of the Cosmos?

The recent discovery of a planet only a few times heavier than the Earth in orbit around a nearby star,* together with the discovery over the past few years of more than 200 Jupiter-like giant planets orbiting other stars, has fuelled interest in the possibility of finding intelligent life elsewhere in the Universe. Many people hope that the fact that there are other 'solar systems' out there must mean that there are other earths – and if there are other earths, surely there must be other people? This is a false argument. In the first place, it seems likely that Earth-like planets are rare. But even if other earths were common, my view is that while life itself may be common, the kind of intelligent, technological civilization that has emerged on Earth may be unique, at least in our Milky Way Galaxy.

* And several more 'other earths' while this book was in press!

PREFACE: THE ONLY INTELLIGENT PLANET

I agree with Lovelock that other Gaias may be relatively common in the Cosmos. But I have reached the conclusion that our kind of intelligent life is so rare that it may be unique to our planet. At the present moment of cosmic time intelligent life exists only on Earth – in the language of Gaia, Earth is the only intelligent planet, at least in our Galaxy.

Whether or not you see the hand of God in any of this, it would mean that we are the most technologically advanced civilization in the Universe, and the only witnesses with an understanding of the origin and nature of the Universe itself. If humankind and Gaia can survive the present crises, the whole of the Milky Way Galaxy may become our home. If not, the death of Gaia may be an event of literally universal significance.

I restrict the discussion to the Milky Way not only because it is our own astronomical backyard, an island in space beyond which we cannot hope to explore physically on any reasonable timescale, but because it is possible that the Universe beyond the Milky Way is infinite. In an infinite Universe, anything is possible, but anything interesting may only be happening infinitely far away from us. The Milky Way contains a few hundred billion stars, but almost certainly contains only one intelligent civilization. In that sense, our civilization is alone, and special. This book tells you why.

Introduction
One in a Trillion

The Universe is big. We live in an expanding bubble of space that burst out from a superhot, superdense state, the Big Bang, 13.7 billion years ago. Because this bubble has only been in existence for 13.7 billion years, the farthest we can see in any direction, even in principle, is the distance that light has travelled in 13.7 billion years. Logically enough, this is 13.7 billion light years, where one light year is the distance light can travel in a year – roughly 9.5 billion km or 5.9 billion miles. So the observable Universe is a bubble centred on the Earth, with a diameter of 27.4 billion light years – a bubble growing in size at a rate of two light years (one on each side) every year.*

This does not mean that we are at the centre of the Universe, any more than a sailor out of sight of land is at the centre of the ocean; the sailor is just at the centre of the circle formed by their own horizon. Sailors in other parts of the ocean are each at the centre of their own 'observable world', surrounded by a horizon. The Universe undoubtedly extends beyond our cosmic horizon, just as the sea extends beyond the sailor's horizon, and may well (unlike the ocean) be infinite. But the puzzle of what lies beyond the cosmic horizon is not one I am going to discuss here.

Within the bubble of the visible Universe, bright stars, more or less like our Sun, are grouped together in islands called galaxies. On the basis of observations made with instruments such as the Hubble Space Telescope, it is estimated that there are hundreds of billions, and perhaps trillions, of galaxies in the observable Universe. All of the

---

* This is a slight oversimplification that does not take account of the expansion of the Universe. I hope my cosmological colleagues will forgive me.

stars we see with our naked eyes are part of one of these islands in space, our home galaxy, called the Milky Way, or simply the Galaxy. But our eyes are hopelessly inadequate for revealing the true nature of the Milky Way Galaxy. In very round numbers, there are roughly as many stars in our Galaxy as there are galaxies in the observable Universe. When I started out in astronomy, back in the 1960s, the round number most often quoted was a hundred billion stars in the Milky Way; as time passed and observations improved, the estimate was increased to a couple of hundred billion, then 'several' hundred billion. Since our telescopes keep getting better, and our observations keep revealing new things, it doesn't seem unreasonable to round this off and say that, very roughly, our Galaxy contains a trillion stars. That means that the Sun is one in a trillion, and our Galaxy is also one in a trillion. And as far as we can tell, the Sun is a pretty ordinary star (although it may have some significant minor peculiarities that I discuss later).

ACROSS THE MILKY WAY

This island of a trillion stars is the background to my story of the emergence of intelligent life on Earth, and the puzzle of whether there is other intelligent life in the Universe. At this stage of human understanding of the Cosmos, all we can reasonably do when seeking an explanation for the emergence of intelligence is to look at the history and geography of our own Galaxy and try to understand why, and how, intelligence has emerged on Earth, and what that tells us about the chances of finding civilizations on other 'earths' somewhere in the Milky Way. If there are enough stars and planets in the entire Universe, there is bound to be another earth out there somewhere; but is there one that is home to another civilization somewhere in our own cosmic back yard?

In those terms, although our Galaxy contains many components, what we are interested in is planets more or less like the Earth that are orbiting stars more or less like the Sun. The Sun and the planets of the Solar System formed together from a collapsing cloud of gas and dust in space a little more than 4.5 billion years ago, when the Universe was only two thirds of its present age. The fact that it took so long before the Solar System formed is not entirely a coincidence. There is overwhelming evidence that the only elements that were produced in any significant quantity in the Big Bang were hydrogen and helium. Heavier elements have been built up since then inside stars, in a process called stellar nucleosynthesis, and scattered across space when those stars die. So there had to be time for several generations of stars to be born, live and die before there were interstellar clouds containing a rich enough mixture of elements such as silicon, oxygen, carbon and nitrogen to make a planet like the Earth.

Part of that process of nucleosynthesis is what keeps a star like the Sun shining. In the heart of the Sun, the extreme temperature and pressure forces nuclei of hydrogen atoms together to fuse them to make nuclei of helium, with energy being released in the process. In other stars, at different stages in their life cycle, helium nuclei are combined to make carbon, oxygen, and so on. All of this activity goes on in a disc of stars, gas and dust that spans a diameter of about 100,000 light years. By measuring the distribution of this material as best we can from inside the Galaxy, and comparing what we see with observations of other galaxies that we can see from the outside, astronomers have found that our Galaxy has a spiral structure, with bands of bright, young stars (called spiral arms) twining outwards from the centre of the disc. It used to be thought that this marked a clean spiral pattern, with four main arms and a few smaller arcs; but recent observations suggest that the pattern is more messy, with spurs sticking out from some of the main arms, fragments of other arms, and even a bar of arms across the middle of the Galaxy.

Nobody knows exactly how the spiral pattern is produced, but it is a common feature of galaxies. The best guess is that it is a density wave, in which stars and gas clouds orbiting around the centre of the Milky Way pile up at certain places, in much the same way that traffic on a motorway piles up in a moving traffic jam near a large, slow-moving load. Gas clouds that get caught up in the rolling traffic jam are squeezed, and some of them collapse to form new stars, which are what makes the spiral pattern stand out. But individual stars are much smaller than these clouds, and pass through the density wave without being affected, on their approximately circular paths around the centre of the Galaxy.

Roughly speaking, the distribution of stars across the Galaxy can be described in terms of four parts. Most of the stars, including the Sun, are concentrated in a thin disc, about 1,000 light years deep, which widens out into a bulge around the centre of the Milky Way. The overall appearance is like two fried eggs stuck back to back. Together, the thin disc and central bulge contain about 90 per cent of the stars in our Galaxy. The thin disc is embedded in a thicker, but sparser, disc of stars, which is itself surrounded by a spherical halo, at least 300,000 light years across, across which a few dozen tightly packed clusters of stars are scattered. As far as bright stars are concerned, that's it. There are, though, two other components, revealed by their gravitational influence on the bright stuff, which are fascinating in their own right but which have no relevance to the search for other earths. The first is a massive black hole right at the centre of the Milky Way, with a diameter twenty times bigger than the distance from the Earth to the Moon, containing several million times as much mass as our Sun. Although that sounds impressive, it is only a few millionths of the mass of all the bright stars in the Galaxy put together. At the other extreme, all of the bright components of the Galaxy are embedded in a cloud of dark material, which holds the Milky Way in its gravitational grip. This material fills a sphere several hundred thousand light years across, and is thought to be made of a cloud of tiny particles, on the scale of atoms. But the total mass of all those tiny particles adds up to ten times as much matter as in all the bright stars in the Galaxy put together.

The search for other planets has so far extended only a tiny way out from the Sun across the thin disc of the Milky Way. At the end of the twentieth century, the technology became just good enough to look for evidence of the influence of planets orbiting nearby stars, and the search has since been pushed out (by no means uniformly) to a distance of a few hundred light years – roughly 0.1 per cent of the diameter of the disc. But the good news is that everywhere we look we find planets. On this evidence, at least half of all the stars we can see probably have planets orbiting around them.

HOT JUPITERS

As yet, the evidence for planets beyond the Solar System is, in almost every case, indirect. Except for a very few cases, we cannot actually see or photograph them yet, and even in those cases the images are only fuzzy blobs; but we detect their influence on their parent stars. As a planet orbits around its star, the gravitational influence of the planet makes the star wobble to and fro by a tiny amount, and this wobble can be detected by studying the spectrum of light from the star. When the star is moving towards us, features in the spectrum shift by a tiny amount towards the blue end of the spectrum; when it is moving away there is a shift towards the red end of the spectrum. This is an example of the Doppler effect, one of the most useful tools in astronomy; the size of the speeds measured in this way, at distances of scores of light years, is about the same as that of an Olympic sprinter. The easiest planets to detect using this method are those that have the biggest influence on their parent star, which means big planets that are close to their star. So it is hardly surprising that most of the several hundred planets discovered so far are indeed big and close to their star. As time passes, however, smaller planets, and ones farther out from their stars, are also being discovered.

As it happens, the very first 'extrasolar' planets to be discovered were something out of the ordinary. But, as the first, they deserve pride of place. The star they orbit around is nothing like the Sun. It is an object called a neutron star, with the prosaic name PSR B1257+12. The PSR stands for pulsar, and the numbers are the coordinates of the position of the object on the sky – the celestial equivalent of a map reference. A pulsar forms when a star much bigger than our Sun reaches the end of its life. The outer layers of the star are blown away in an explosion called a supernova, and the inner region collapses into a ball of neutrons (hence the name) only about 10 km across but containing roughly as much mass as our Sun (330,000 times the mass of the Earth). The density of a neutron star is the same as the density of an atomic nucleus, and when they are first formed they are spinning very rapidly and have intense magnetic fields. This combination produces a beam of radio noise, which sweeps around the sky like the beam of a cosmic lighthouse; if the beam happens to be lined up so that it sweeps across the Earth, our radio telescopes detect this as a regular ticking. It is these 'pulses' of radio noise that give pulsars their name. Some 'tick' once every few milliseconds, and PSR B1257+12 is just such a pulsar. In 1992, Alex Wolszczan and Dale Frail, of Penn State University, measured tiny changes in the rate at which this pulsar is ticking, and explained these variations as due to the effect of two planets orbiting around the dying star. A couple of years later they reported the discovery of a third planet in the system. These planets have masses equivalent to 4.3 times, 3.9 times and one fiftieth of the mass of the Earth, and orbit the star once every 67 days, once every 98 days and once every 25 days. In 2005, Wolszczan and his colleague Maciej Konacki announced that they had identified a fourth planet in the same system, a tiny object roughly one tenth of the mass of the dwarf planet Pluto (only 0.04 per cent of the mass of the Earth) that takes 1,250 days to orbit once around the pulsar. At least one other pulsar, PSR B1620-26, is now also known to have one or more planetary companions.

These planets cannot possibly have survived the supernova explosion in which the pulsar was formed. Any planets that orbited the original star must have been destroyed in the blast. So they must have formed from the cloud of debris left around the neutron star in the aftermath of the explosion. This was the first direct proof that planets can form from clouds of debris around a star, rather than by some more exotic process such as a collision, or close encounter, involving two stars. Since stellar close encounters are rare, but all stars form from collapsing clouds of interstellar material, the implication was that planets ought to be common companions to stars. And that was soon borne out by more observations.

The first confirmed discovery of a planet orbiting a star similar to the Sun was made in 1995. The star is 51 Pegasi (known for short as 51 Peg), and the discovery was made by two Swiss astronomers,

Michel Mayor and Didier Queloz, using the Doppler technique. The surprise was that the discovery was almost too easy. It was easier than had been expected because the planet is big and orbits very close to its parent star, a combination which produces a strong Doppler 'signal'. In our Solar System, there are four small, rocky planets orbiting close to the Sun, and four very large, gaseous planets farther out from the Sun. Distances are measured in terms of the astronomical unit, or AU, which is equivalent to the average distance from the Earth to the Sun. Masses are measured in terms of the mass of the Earth. The small planet Mercury, closest to the Sun at a distance of 0.39 AU, has a mass 5 per cent that of the Earth, but the largest planet in the Solar System, Jupiter, has a mass more than 300 times that of the Earth, about 0.1 per cent of the mass of the Sun, which it orbits at a distance of 5.2 AU. The planet found around 51 Peg has more than half as much mass as Jupiter (three thousand times the mass of Mercury), but orbits its star at a distance of only 0.05 AU (only a little more than a tenth of the distance from Mercury to the Sun).

Nobody had expected to find such 'hot jupiters', as they soon became known, because large gaseous planets cannot form close to their parent star. So they must have migrated inwards from the orbits where they were born. This has significant implications for the search for intelligent life in the Universe; but the first importance of this discovery, and those that soon followed, was (and is) that planets are common.

When the first extrasolar planets were discovered, the news was so exciting that each new discovery rated a separate scientific paper in a prestigious, rapid-publication scientific journal such as the weekly *Nature*, and often made headlines in the general media. More than a decade down the line, at the time of writing nearly four hundred extrasolar planets are known, and 'new' ones are being discovered at the rate of a dozen or so every year. But the news seldom makes headlines even in the scientific journals, let alone the popular press, unless there is something special about the latest discovery – especially if the planet is particularly Earth-like, or is in an Earth-like orbit around its parent star. The whole story, of the number and variety of extrasolar planets, or exoplanets, is much greater than the sum of its parts.

Part of that story concerns the variety of different techniques used to discover planets orbiting other stars. As well as the tried and tested Doppler method, some planets have been detected by their effect on the light from their star as they pass in front of it, in a kind of mini-eclipse called a transit. Others have been located by their gravitational influence on the very discs of dusty material around stars in which planets form, in a pleasing confirmation of our ideas about planetary formation. A very few extrasolar planets have now been photographed directly, as tiny specks in images obtained using infrared telescopes. And there are other techniques. Almost all these discoveries, though, have one thing in common – so far, they have been successful using ground-based telescopes. But there are now space observatories dedicated to the search for extrasolar planets – including Kepler, launched in 2009 – and the number of discoveries is increasing dramatically as they make observations from above the obscuring layers of the Earth's atmosphere. So this is a particularly good time to take stock of the first phase of exoplanet observations.

PLANETS IN PROFUSION

Because this early phase of the search for exoplanets inevitably picked up a lot of large planets in extreme orbits, from the point of view of the search for other earths it's particularly encouraging that more than twenty of the extrasolar planetary systems now known contain more than one planet – nearly a hundred planets are known to be orbiting normal stars in multiple planetary systems. Even if we are still mostly discovering large planets ('jupiters'), this suggests that such planets do not form in isolation, and at least some of those jupiters are accompanied by smaller, rocky planets ('earths'), as is the case in our Solar System.

Most of the planets discovered so far orbit around stars that are roughly similar to our Sun (and which, for historical reasons, are known as F, G or K stars – the Sun is a yellow G2 dwarf star, in this classification). This is partly because the astronomers involved in the search are, naturally, particularly interested in such stars. But there are also reasons to think, as I discuss later, that stars much bigger or much smaller than the Sun are unlikely to provide the conditions suitable for the formation of other earths. Although the vast majority of planets discovered so far are at least ten times as massive as the Earth, and some are much bigger than Jupiter, the fact that even a few planets with masses only a few times that of the Earth have been found, in spite of the difficulties of detecting them, suggests that such rocky planets are actually quite common.

Unfortunately, many of these roughly Earth-sized planets are in tight orbits around their parent stars, like the orbits of the hot jupiters. The natural explanation is that they are the leftover rocky cores of hot jupiters that have evaporated away in the heat from their stars. A typical example is the planet COROT-7b, which circles its parent star at a distance of only 2.6 million km, in just over 20 hours! The star, COROT-7, is a yellow dwarf, only about 1.5 billion years old, similar to, but slightly smaller and cooler than, the Sun; COROT-7b is less than twice the diameter of the Earth, with a density similar to that of our planet. But the planet is so close to the star that the surface temperature is likely to exceed 2,000 °C, hot enough to melt rock. In any case, because its orbit is not quite circular and it is so close to its star, tidal forces are rhythmically squeezing the planet, heating its interior and almost certainly producing continuous volcanic activity at the surface. Not a likely home for life. Computer simulations suggest that COROT-7b started life as a gas giant with about the same mass as Saturn (roughly a hundred times the mass of the Earth) in an orbit 50 per cent larger than its present one.

After the initial surprising (but easy) discovery of hot jupiters, it has emerged that most 'jupiters' (a term which embraces all large, gaseous planets, in the same way that the term 'earths' embraces all small, rocky planets) are actually in orbits much farther out from their parent star, and that planetary systems with one or two such planets in orbits roughly similar to the orbits of Jupiter and Saturn in our own Solar System are not uncommon. Perhaps they are accompanied by other earths; computer simulations of the way planets form around stars certainly suggest that giants are inevitably accompanied by smaller, rocky planets. Some astronomers now think that every Sun-like star has at least one 'Earth like' planet – although by the term 'Earth like' they include planets like Venus and Mars, not solely true other earths – in the so-called 'habitable zone' around a star, the region where liquid water can exist. One big difference, though, is that the orbits of the Jupiter-like exoplanets mostly seem to be more elliptical (less circular) than the orbits of the equivalent planets in the Solar System.

Two other discoveries have stimulated speculation about the possibility of life elsewhere in the Universe. One was the discovery of a planet only slightly bigger than the Earth, although as it happens it orbits a much fainter star than our Sun. The star is a dim red dwarf, known as Gliese 581, just one third of the mass of the Sun and only a little more than 20 light years away from us in the direction of the constellation Libra. The planet is only 1.9 times more massive than the Earth, and orbits the red dwarf once every 3.15 days, too close for liquid water to exist on its surface. But another planet in the same system has a mass of only seven Earth masses, and orbits farther out, once every 66.8 days. It is smack in the life zone for a red dwarf, where temperatures are between about 0 °C and 40 °C. It could be covered in water. Red dwarf stars are common in our celestial neighbourhood, and like Gliese 581 they seem to provide homes for 'super earths' rather than jupiters. This makes them prime targets for future exoplanet studies, even though they are much smaller than the Sun.

Water lies at the heart of the other recent discovery. In 2007 a large team of astronomers reported that they had detected water in the atmosphere of one of the hot jupiters. They were able to do so because the planet, which had already been detected in a previous survey, passed in front of its parent star, HD 189733, which lies 64 light years from us in the direction of the constellation known as the Fox. Some of the light from the star passed through the atmosphere of the giant planet before carrying on to be picked up by the Spitzer Space Telescope, operating at infrared wavelengths. The effect of water in the atmosphere of the planet showed up clearly in the infrared signature of the light.

As if all this wasn't intriguing enough, in 2008 scientists from the University of Toronto, using the Gemini Observatory in Hawaii, obtained the first picture of a planet in orbit around a Sun-like star.

INTRODUCTION: ONE IN A TRILLION

Putting everything together, we know that planets are common; we know that rocky planets exist in the life zones around stars; and we know that there is water on worlds orbiting other stars. Out of a trillion stars in the Milky Way Galaxy, a conservative estimate would seem to be that a billion of them (just 0.1 per cent) are orbited by what we might call 'wet earths'. How could life get a grip on such a potential planetary home?

DUSTY BEGINNINGS

Everything we know about the way planets form suggests that life is an almost inevitable by-product of the formation of a wet earth. Stars – and planets – form from collapsing clouds of gas and dust in space. There is plenty of this material around in the Milky Way, although it needs a trigger – a squeeze of some kind – before it will start collapsing. But the dust, which is essential for the formation of Earth-like planets, is only a small component of the total. The clouds between the stars are 98 per cent gas, almost all of it hydrogen and helium left over from the Big Bang, which leaves just 2 per cent for everything else. We can see young, newly formed stars in clouds such as the Orion Nebula, which is about 25 light years across and 1,400 light years away; it contains thousands of very young stars, hundreds of which are so young that they have not settled down into a mature state like the Sun. Because the Orion Nebula is the closest star-forming region to the Earth, some of these objects can be studied in great detail, and in more than 150 cases discs of dusty material have been identified, and some even photographed, around young stars in the nebula. Similar discs have also been identified associated with young stars in other parts of our neighbourhood.

These discs are the sites where new planets are forming. In some cases, planets have been detected because of their gravitational effects on the discs, warping them, or clearing the dust from some regions. So we know planets are associated with these discs, which are therefore called proto-planetary discs, or PPDs. A lot of information about PPDs comes from studies at infrared wavelengths, because the discs are much cooler than stars – a cool object radiates most of its energy at longer wavelengths than a hot object. The discs are warmed by the energy from their parent stars, which is at shorter wavelengths, and re-radiate the energy in the infrared. The nature of the infrared radiation reveals that the size of the dust grains involved in this process is tiny – up to about 10 microns, or ten millionths of a metre, in diameter. That is roughly the size of a bacterium. But the discs are almost entirely made of dust, because any gas that was in the original cloud but didn't collapse into the central star got blown out of the system by the young star's radiation.

In one well-studied system, about 53 light years away and known as Beta Pictoris, there is a few hundred times the mass of the Earth (about the same as the mass of Jupiter) present in the form of dust in the disc. A combination of optical and infrared studies shows that there are denser concentrations of dust in rings within the disc, which may be the sites of planet formation. A great deal of the dust we now see there may actually be a result of collisions between the first rocky objects formed in the system, called planetisimals, which collide with one another and eventually build up into large planets.

In this sense, Beta Pictoris, although it has an estimated age of only a few tens of millions of years, is a relatively old system. The even younger star HL Tauri has a disc containing about a tenth as much mass as our Sun – ten times the mass of all the planets in our Solar System put together – extending over a diameter of 2,000 AU. That probably is primordial cosmic dust, not the secondary dust seen around Beta Pictoris. Some 60 per cent of stars younger than 400 million years have dusty discs associated with them, but only 9 per cent of stars older than 400 million years possess such discs. All the evidence is that the discs have gone because the dust has been swept up in the formation of planets, and the timescale involved, a few hundred million years, closely agrees with the timescale for the formation of the Solar System, inferred from a variety of studies (more of this later).

But what does the dust contain? One of the most important ingredients is water, in the form of ice. The most common element in the Universe is hydrogen, about 73 per cent by weight. The next most common is helium, about 25 per cent. Both of these were produced in the Big Bang, but helium does not react chemically so it plays no direct part in life processes. The third most common element is oxygen, at 0.73 per cent, followed by carbon, at 0.29 per cent. In terms of mass, the next most abundant element is iron; but in terms of the number of atoms around, the fifth spot is taken by nitrogen. With a somewhat cavalier attitude towards chemical subtleties, astronomers lump all the elements except hydrogen and helium together under the name 'metals'. But whatever the name, the important point is that oxygen is the most common reactive element after hydrogen, and hydrogen and oxygen react together eagerly to make water. So interstellar grains and grains in PPDs are bound to contain a lot of water ice, which forms a layer on the surface of any solid particles, such as grains of carbon (graphite).

These grains are about as substantial as the solid particles in cigarette smoke. Because of the way the ice forms in space, straight from the vapour to the solid without passing through a water phase, it resembles snowflakes rather than ice cubes, and when one 'snowflake' bumps into another it will tend to stick, building up larger grains. The stickiness of the grains is helped because water molecules have a positive electric charge at one end and a negative charge at the other end, producing electric forces which hold the grains together like tiny bar magnets. Soon, by astronomical standards, the grains in a PPD are big enough to pull on each other by gravity, and the process of planet formation begins.

The kinds of molecules that can form on the surfaces of such grains are surprisingly complex. We know of the existence of complex molecules in space from observations of the large clouds of gas and dust between the stars, made at radio wavelengths. Just as the atoms of an individual element produce a characteristic pattern like a bar code in the spectrum of visible light, so different kinds of complex molecules each produce their own distinctive pattern in the radio part of the spectrum. More than 140 different kinds of molecule have now been identified in space in this way. These range from such simple and unsurprising substances as molecular hydrogen (H_2) and water (H_2O) up to compounds containing ten or more atoms in each molecule, such as n-propyl cyanide (C_3H_7CN) and ethyl formate

(C_2H_5OCHO). As with these two examples, very many of these complex molecules are made from combinations of atoms of carbon, hydrogen, oxygen and nitrogen. At one level, this is not surprising, since as I have mentioned these are the four most common reactive elements in the Universe, in terms of the numbers of atoms around. At another level, it is deeply significant because these four elements, collectively dubbed 'CHON', are the most important elements in the chemistry of life. Clearly, that has to do with the fact that life utilizes the chemical materials that are available. But it also has to do with the unusual chemical properties of carbon.

COSMIC CHEMISTRY

Carbon atoms have an unusual ability to combine strongly with up to four other atoms at a time, including other atoms of carbon. The simplest way to picture this is to imagine that a carbon atom has four hooks sticking out from its surface, and each of these can latch on to another atom to make a chemical bond. In the simplest example, each molecule of the compound methane is made of a single carbon atom surrounded by four hydrogen atoms, which are attached to it by bonds – CH_4. But carbon atoms can also link up with one another fore and aft to form chains, linking each carbon atom in the chain with two other carbon atoms, but leaving two bonds free to hook up with other kinds of atoms, and leaving the two carbon atoms at the ends of the chain each with three spare bonds. Or the chain may become a ring, with carbon atoms forming a closed loop, still with two bonds available for each atom in the ring to form other linkages. Even complex carbon-based molecules, including other rings and chains, can attach to other carbon chains or to other rings. It is this rich potential for carbon chemistry which makes the complexity of life possible. Indeed, when chemists first began to study the complexity of life, and realized that it involves carbon so intimately, the term 'organic chemistry' became synonymous with 'carbon chemistry'.

There are two key components of the chemistry of life. To non-biologists, the most widely known life molecule is DNA, or deoxyribonucleic acid. This is the molecule within the cells of living things, including ourselves, which carries the genetic code. The genetic code contains the instructions, rather like a recipe, which tell a fertilized cell how to develop and grow into an adult. But it also contains the instructions which enable each cell to operate in the right way to keep the adult organism functioning – how to be a liver cell, for example, or how to absorb oxygen in the lungs. The mechanism of the cell also involves another molecule, ribonucleic acid, or RNA. As the name suggests, molecules of DNA are essentially the same as molecules of RNA but with oxygen atoms removed.

The 'ribo' part of the name comes from 'ribose' (strictly speaking, the names should be ribosenucleic acid and deoxyribosenucleic acid). Ribose ($C_5H_{10}O_5$) is a simple sugar, but it lies at the heart of DNA and RNA. Each molecule of ribose is made of a core of four carbon atoms and one oxygen atom linked in a pentagonal shape. Each of the four carbon atoms in the pentagon has two spare bonds with which to link up with other atoms or molecules. In ribose itself, these attachments link the pentagon to hydrogen atoms, oxygen atoms, and one more carbon atom, making five in all, which is itself joined to more hydrogen and oxygen; but any of these attachments can be replaced by other links, including links to complex groups which themselves link up with other rings or chains. In DNA and RNA, each sugar ring is attached to a complex known as the phosphate group, which is itself attached to another sugar ring. So the basic structure of both of the life molecules is a chain, or spine, of alternating sugar and phosphate groups, with interesting things sticking out from the spine. It is the interesting things that carry the code of life, spelling out the message in what is in effect a four-letter alphabet with each letter corresponding to a different chemical group. But that is not a story to go into here; from the point of view of interstellar chemistry, it is the basic building blocks of DNA, the ribose molecules, that are significant.

Nobody has yet detected ribose in space. But astronomers have detected the spectroscopic signature of a simpler sugar called glycolaldehyde. Glycolaldehyde is made up of two carbon atoms, two oxygen atoms and four hydrogen atoms (usually written as $H_2COHCHO$, which reflects the structure of the molecule), and is known, logically enough, as a '2-carbon sugar'. Glycolaldehyde readily combines, under conditions simulating those in interstellar clouds, with a 3-carbon sugar, making the 5-carbon sugar ribose. We have not yet found the building blocks of DNA in space; but we have found the building blocks of the building blocks.

The other kind of 'life molecule' is protein. Proteins are the structural material of the body; they always contain atoms of carbon, hydrogen, oxygen and nitrogen, often sulphur, and some contain phosphorus. Things like hair and muscle are made of proteins in the form of long chains, not unlike the long chains of sugar and phosphate in DNA and RNA molecules; things like the haemoglobin that carries oxygen around in your blood are forms of protein in which the chains are curled up into little balls. Other globular proteins act as enzymes, which encourage certain chemical reactions that are beneficial to life, or inhibit chemical reactions that are detrimental to life. There is such a variety of proteins because they are built up from a wide variety of sub-units, called amino acids.

Amino acid molecules typically have weights corresponding to a hundred or so units on the standard scale where the weight of a carbon atom is defined as 12, but the weights of protein molecules range from a few thousand units to a few million units on the same scale, which gives you a rough idea of how many amino acid units it takes to make a protein molecule. One way of looking at this is that half of the mass of all the biological material on Earth is in the form of amino acids. But even though a specific protein molecule may contain tens of thousands, or hundreds of thousands, of separate amino acid units, all the proteins found in all the forms of life on Earth are made from combinations of just twenty different amino acids. In the same way, every word in the English language is made up from different combinations of just 26 sub-units, the letters of the alphabet. There are many other kinds of amino acid, but they are not used to make protein by life as we know it.

If a chemist wishes to synthesize amino acids in the laboratory, it is relatively easy and quick to do so by starting out with compounds such as formaldehyde (HCHO), methanol (CH_3OH) and formamide

INTRODUCTION: ONE IN A TRILLION

($HCONH_2$), all of which will be to hand in any well-stocked chemical lab. With such materials readily available, it would be crazy to start out from the basics – water, nitrogen and carbon dioxide. But the chemistry lab isn't the only place you will find such compounds. One of the most dramatic results of the investigation of molecular clouds is the discovery that all of the compounds used routinely in the lab to synthesize amino acids (including the three just mentioned) are found in space, together with others such as ethyl formate (C_2H_5OCHO) and n-propyl cyanide (C_3H_7CN). In a sense, the molecular clouds are well-stocked chemical laboratories, where complex molecules are built up not atom by atom, but by joining together slightly less complex sub-units.

There have also been claims that the simplest amino acid, glycine (H_2NH_2CCOOH), has been detected in space. It is very difficult to pick out the spectroscopic signature of such a complex molecule, let alone those of more complex amino acids, and these claims have not been universally accepted by astronomers, even though amino acids have been found in rocks from space left over from the formation of the Solar System, which occasionally fall to Earth as meteorites. The claims have been bolstered, though, by the recent detection in space of amino acetonitrile (NH_2CH_2CN), which is regarded as a chemical precursor of glycine. But even if we take the cautious view and leave these claims to one side, that still means that, echoing the situation with DNA, with the identification of compounds such as formaldehyde, methanol and formamide, although we have not yet found the building blocks of protein in space, we have found the building blocks of the building blocks.

Complex organic molecules can only be built up in the molecular clouds because those clouds contain dust as well as gas. If all the material in the clouds were in the form of gas, even if by some unimaginable process a complex molecule such as NH_2CH_2CN did exist, how could it grow? You might imagine that a collision with a molecule of oxygen, O_2, would provide an opportunity to capture some of the additional atoms needed to make glycine, H_2NH_2CCOOH. But the impact of the oxygen molecule would be more likely to break the amino acetonitrile apart, rather than to encourage it to grow. But tiny solid grains, coated with a snowy layer of ice (not just water ice, but also things like frozen methane and ammonia), provide sites where molecules can stick and be held alongside each other for long enough for the appropriate chemical reactions to take place.

Old stars swell up near the end of their lives, and eject material out into space. Spectroscopic studies show that this material includes grains of solid carbon, silicates and silicon carbide (SiC), which is the most common solid component definitely identified in the dust around stars, although there are many as yet unidentified spectral features as well. Laboratory experiments simulating the conditions on the surfaces of such particles in space have confirmed that they provide places where the kinds of chemical reactions needed to make the kinds of complex organic molecules we detect in space can take place. Some of these studies suggest that the grains may not simply provide a surface where the reactions can occur, but that there may be chemical bonds between the molecules and the surface itself. That would explain how the molecules stick around for long enough for the reactions to take place even in relatively warm parts of a molecular cloud. As long as they do stick, there is plenty of time for the reactions to happen, because molecular clouds may wander around the Galaxy for millions – even billions – of years before part of the cloud collapses to form a group of new stars. When the grains are warmed by the heat from a newly forming star, the complex molecules can be liberated and spread through the molecular cloud, where they can be detected by our radio telescopes.

In this context, it is almost an anticlimax, but still significant, that a simple organic molecule, methane, was detected in the atmosphere of one of the hot jupiters in 2008. This was no surprise – methane is an important component of the atmosphere of Jupiter itself. But it was still regarded as a landmark event. For the record, the planet is the same one where water was identified earlier, orbiting the star HD 189733. Astronomers working with the Spitzer Space Telescope have also found large amounts of hydrogen cyanide, acetylene, carbon dioxide and water vapour in the discs around young stars where planets form. And a team from the Carnegie Institution used the Hubble Space Telescope to analyse light from a star known as HR 4796A,

270 light years away in the direction of the constellation Centaurus, to determine that the red colour of the dusty disc around the star is caused by the presence of organic compounds known as tholins. Tholins are large, complex organic molecules that are manufactured by the action of ultraviolet light on simpler compounds such as methane, ammonia and water vapour. They can be synthesized in the lab, but do not occur naturally on Earth today because they would be destroyed by reacting with oxygen in the atmosphere as fast as they formed. But their presence explains the reddish-brown hue of Saturn's moon Titan, they are present in comets and on asteroids, and they may well have been present on Earth when it was young. Tholins are widely regarded as precursors of life on Earth, which made their discovery in the disc around HR 4796A hot news.

This is not the same, though, as finding such compounds on a planet. When planets like the Earth form by the accretion of larger and larger lumps of rock, they get hot, because of the kinetic energy released by all those rocks smashing together. A rocky planet starts its life in a sterile, molten state, certainly hot enough to destroy any organic molecules present in the material from which it formed. The importance of all the observations of organic material in space is that they tell us that there is a great reservoir of such material available to fall down on to the planets after they are cool enough for the complex molecules to survive. Life does not have to be 'invented' from scratch on each new planet from the basics of water, carbon dioxide and nitrogen, any more than an organic chemist has to synthesize amino acids from the basics of water, carbon dioxide and nitrogen.

THE LIFE OF GAIA

In the words of James Lovelock, the founder of Gaia theory and someone who has thought more deeply than most people about the nature and meaning of life, 'It seems almost as if our Galaxy were a giant warehouse containing the spare parts needed for life.' Before looking at just how quickly life can get a grip on a planet like the Earth, it's worth looking briefly at Lovelock's Big Idea, Gaia, since the search for 'other earths' is in many ways a search for 'other Gaias', and NASA's plans for the detection of other Earth-like planets very much depend on the understanding of the relationship between life and the Universe developed by Lovelock in the context of Gaia theory.

Ironically, when Lovelock first came up with the key concept it was most unwelcome at NASA, for whom he was working as a consultant. In 1965, his role with NASA was to help with the design of experiments to search for evidence of life on Mars. Detectors were to have been sent to Mars by an unmanned probe, although as it happens this particular project was cancelled before launch. While his colleagues concentrated on ideas for experiments to test for the presence of specific life forms similar to those on Earth, Lovelock became intrigued by the idea of devising a general test for the presence of life on a planetary scale. He reasoned that one of the key features of life is that it maintains itself and its surroundings in a state far away from chemical equilibrium. At a personal level, your body, for example, is warmer than its surroundings. On a global scale, the obvious example is the presence of a large amount of oxygen in the atmosphere of the Earth. Oxygen is very reactive, and if there were no life on Earth it would soon get locked up in compounds such as water, carbon dioxide and oxides of nitrogen. It is life, using energy from the Sun to drive the chemical processes of life, that puts oxygen back into the air as fast as it is used up.

At that time, in the spring of 1965, nobody knew what the atmosphere of Mars was made of. Lovelock suggested that a better way to search for life there, rather than going to all the trouble of sending a probe, would be to build a telescope able to scan the spectrum of Mars in the infrared and find out what gases were in its atmosphere. If they were inactive gases like carbon dioxide, that would prove that Mars is a dead planet now, whatever may have happened to it in the past.

In September 1965, unaware of Lovelock's suggestion, a team of French astronomers did indeed study the infrared spectrum of the atmosphere of Mars, and found that it is almost entirely carbon dioxide, in chemical equilibrium. The implications were clear, at least to

Lovelock. There was no life on Mars, and it was pointless sending instruments to look for it. Although, of course, there might be other interesting things to study on Mars, carrying life-detection experiments on the probes was a waste of valuable resources. This conclusion did not go down too well with scientists who had devoted their careers to designing such detectors, and even now, more than forty years later, NASA is still sending probes to look for life on Mars.

When the atmosphere of Venus also turned out to be dominated by carbon dioxide, Lovelock began to think about what it is that makes the Earth special. In Lovelock's words, 'the air we breathe can only be an artefact, maintained in a steady state far from chemical equilibrium by biological processes.' Living things, he concluded, must be regulating the composition of the atmosphere, not just today but throughout the history of life on Earth – literally for billions of years.

This resolved a puzzle known to astronomers at the time as the 'faint young Sun paradox'. Astronomers already knew in the 1960s, from their calculations of how stars work and their studies of other stars, that when the Sun was young it was about 25 per cent cooler than it is today. Other things being equal, the Earth would have frozen. The resolution of the puzzle is that when the Earth was young, its atmosphere was rich in carbon dioxide, and a strong greenhouse effect would have kept the surface of the planet warm. But the puzzle then becomes, why didn't a runaway greenhouse effect set in as the Sun warmed, searing the surface of the planet, as seems to have happened on Venus? The answer, Lovelock saw, is that life has regulated the composition of the atmosphere, removing the carbon dioxide gradually as the Sun warmed, keeping temperatures on Earth comfortable for life.

Over many years, this insight led Lovelock to develop a fully fledged theory of how the living and non-living components of the Earth interact through a series of feedback processes to maintain suitable conditions for life – scientists uncomfortable with the name 'Gaia' call this 'Earth System Science'. But there is no consciousness involved, any more than you are conscious of the many feedback processes which keep the temperature of your body more or less constant.

More of Gaia theory – or, if you like, Earth System Science – later.

The relevant point here is that Lovelock's insight gives us the means to search for life beyond the Solar System in exactly the same way that those French astronomers inadvertently proved that there is no life on Mars. And the delicious irony is that NASA scientists are now fully behind Lovelock, since this time their careers depend on using Gaia theory to look for life elsewhere.

SEARCHING FOR OTHER GAIAS

Using the technique to search for evidence of life on extrasolar planets involves an enormous scaling-up operation. Part of the problem is the need to make observations in the infrared part of the spectrum, where the signatures of interesting molecules such as carbon dioxide, water vapour and ozone (the tri-atomic form of oxygen) are strong. Unfortunately, infrared wavelengths of radiation are absorbed by the Earth's atmosphere, and by dust in the inner part of the Solar System, making it difficult to pick out weak features. To study the atmosphere of Mars in the infrared, all you need is a decent telescope on a mountain in France, above most of the obscuring layers of the Earth's atmosphere. But to study the atmospheres of Earth-sized extrasolar planets in the infrared you need a sophisticated space telescope, far out in the orbit of Jupiter, away from all the interference associated with the inner Solar System. Nearly half a century after the landmark French work, we at last have the technology to do the job; all we need is the money to put the technology to use.

Stimulated by the discovery of extrasolar planets in the 1990s, both European and American space agencies (ESA and NASA) came up with detailed plans for such a telescope. The European version was called Darwin, and the American version the Terrestrial Planet Finder, or TPF. Their plans were very similar, and if anything ever gets off the drawing board it will probably be a joint mission with yet another name, but using the same principles. The telescope will operate in the infrared because that is where terrestrial (Earth-like) planets are brightest – the energy they absorb from their parent stars is re-radiated at longer wavelengths because the planets are cooler than the stars. To make the detection of terrestrial planets feasible, you need as big a telescope as possible, but fortunately there is a technique called interferometry, which makes it possible to use several small telescopes linked together to mimic some of the properties of a single, much larger telescope. This involves adding up the light from all the small telescopes in the right way; but it can also be used to subtract the light from one telescope from that of another. Making a virtue out of the necessity of using interferometry, in this way the astronomers have come up with a neat trick whereby all the light from the object at the centre of the field of view is cancelled out, so that in effect the star disappears, leaving a clear view of the much fainter planets orbiting the star.

Early designs for the project involved a group of six infrared telescopes, each with a mirror about 1.5 metres in diameter, flying in formation at the corners of a hexagon. They would be linked to each other and to another unmanned spacecraft at the centre of the formation by laser beams, enabling them to maintain a tight formation while the central spacecraft sent the data back to Earth. The distance between adjacent spacecraft would be about 100 metres, but they would have to maintain their position to an accuracy of less than a millimetre, while orbiting the Sun at a distance of more than 600 million kilometres from Earth. Later versions of the plan have reduced the number of telescopes to four, but increased the diameter of each mirror to around 6 metres; but whatever is finally decided upon the principles remain the same.

Astonishingly, the experts assure us that such a mission could be launched within a few years – certainly less than a decade – of getting the go ahead. But it would then take years of painstaking observations, in a multi-stage process, to locate any other earths in our galactic neighbourhood.

The first step will be to find terrestrial planets. This means planets more or less the size of Earth, Venus and Mars in orbits more or less like those of Earth, Venus and Mars – planets like Mercury are too small and too close to their parent stars to be detected using this technology. Astronomers have drawn up a target list of 200 nearby stars which they regard as good candidates for having planetary systems, within about 50 light years from us. In order simply to detect any terrestrial planets orbiting around those stars, the telescope would need to look at each system for a few tens of hours, so it would take a couple of years to search through this entire catalogue and eliminate the stars without planets. Then, things will get interesting.

The next stage will be to look for planets with atmospheres, and the best signature of this is likely to come from carbon dioxide, which is both a strong absorber and a strong emitter of infrared radiation – that is why it is such a potent greenhouse gas. Identifying the carbon dioxide signature in the spectrum of a planet would take about 200 hours of observing time for each planet, so the best eighty candidates will be studied in this way over the next couple of years. With luck, this phase of the programme will also reveal the presence of water vapour in the atmospheres of some of the planets.

Only then will it be possible, at last, to apply the Lovelock test. The presence of large amounts of a reactive gas like oxygen in the atmosphere of a planet would be proof that it was in a non-equilibrium state, and a probable home to life. Just as on Earth itself, if there is oxygen in the atmosphere of any terrestrial planet orbiting a Sun-like star interactions with the ultraviolet radiation from the star will convert some of the oxygen (O_2) into the tri-atomic form, ozone (O_3), which has a distinctive infrared signature, but much weaker than the carbon dioxide signature. It will take a further couple of years for the telescope to spend some 800 hours at a time observing, in turn, each of the twenty best candidates identified in the previous stage of the search, to look for the presence of ozone.

Putting all of that together, astronomers are confident that if by some magic funding for such a mission became available tomorrow, within twenty years we would know for sure if there were other life-bearing planets – other Gaias – within 50 light years of Earth. That may sound like a daunting timescale for such a project; but I happen to be writing this paragraph on the fortieth anniversary of the first Moon landing. The time from now to the discovery of another earth could be less than half of the time from Apollo 11 to today, so both could happen within my lifetime. Given everything we now know about stars and planets, it will be astonishing if the search, when it does go ahead, proves fruitless. I'm as sure as I can be without data from something like the Darwin/TPF project that there are indeed other Gaias.

But finding life is only part of the quest. Is there intelligent life out there? That's the question this book sets out to answer. So far, the only intelligent civilization we know about is our own, and one thing we do know is that it took a very long time for intelligence to emerge on planet Earth. If (and it's a big if) that is typical, perhaps it is important that life gets a grip early on in the history of a planet. So how did life get started on Earth, and what can that tell us about the chances of finding intelligent life elsewhere? Is our Solar System really one in a trillion? Or are 'they' out there, waiting to hear from us?

I

Two Paradoxes and an Equation

Life got a grip on Earth with almost indecent haste. When our planet was young, it was bombarded by debris left over from the formation of the Solar System, creating a hostile environment in which life could not get a grip. This bombardment also scarred the Moon. Studies of lunar cratering and other evidence tells us that this bombardment tailed off about 3.9 billion years ago, some 600 million years after the formation of the Solar System. But there is evidence that as soon as the bombardment ended, life began.

The oldest evidence is not of life itself, but of a characteristic signature of life. Carbon atoms come in several varieties, called isotopes, which have the same chemical properties but different weights. The most common stable form is called carbon-12, but there is another, slightly heavier, stable form called carbon-13. Living things prefer to take up carbon-12 from their environment, so they produce an excess of the lighter isotope compared with their surroundings. Ancient rocks from Greenland, just over 3.8 billion years old, contain exactly this isotopic signature of life. This suggests that biological processes were going on on Earth as soon as the bombardment ended; the most plausible explanation for this is that the bombardment carried with it the seeds of life, brewed up from the chemical cocktail known to exist in the clouds of gas and dust from which stars form.

Apart from the precursors to life detected in these clouds by spectroscopy, both amino acids and sugars have been found in meteorites and in the dust trails from meteors burning up in the Earth's atmosphere today. Confirmation of the origin of these complex organic molecules has come from NASA experiments in which amino acids were synthesized under conditions mimicking those that exist in dense interstellar clouds. These organic molecules include glycine, alanine and serine, which are basic parts of protein. And in 2009, a team of NASA scientists announced that they had found glycine in material returned to Earth by the spaceprobe Stardust from comet Wild 2. This was the first confirmed detection of an amino acid in space. Such material must have rained down on the young Earth in profusion at the end of the early bombardment; as a spokesman for the Stardust team put it, 'our discovery supports the theory that some of life's ingredients formed in space and were delivered to Earth long ago by meteorite and comet impacts.'

Even so, when we talk about life most people want direct evidence of living creatures – fossils. The earliest known fossils are the remains of colonies of bacteria known as stromatolites. These are found in rocks as old as 3.6 billion years, laid down less than a billion years after the formation of the Solar System, and less than 300 million years after the end of the early bombardment. Stromatolites are not only direct evidence of early life; they are also evidence that by that time there was already a complex ecosystem, with many different kinds of microbes living alongside each other and interacting with one another. Clearly, life itself must have got started even earlier than 3.6 billion years ago, as the isotope evidence implies.

But the stromatolites also highlight one of the most important features of life on Earth. The chemistry of life always takes place in a special environment, protected from the outside world. That special place is the cell, a tiny bag of watery liquid containing all of the requirements for life. Within the cell, DNA and RNA can go about their business of making copies of themselves (reproduction) and providing the instructions for the manufacture of protein molecules. This is the essential chemistry of life. But it has to be contained in a secure environment.

The best way to appreciate the importance of the cell is to look at the role of enzymes, the proteins that encourage the essential chemical reactions of life to take place. Enzymes are not particularly rugged molecules. If they get too hot or too cold, they fall apart. If their surroundings are too acid or too alkaline, they fall apart. If they fall apart, they can no longer do their job, and life stops. So they have to operate inside a protective wall, a special kind of wall which allows some molecules in but keeps others out, and which allows some molecules out but keeps others in. This wall is called a semi-permeable membrane, and it is the wall that surrounds the bubble of a cell. One of the defining characteristics of life – perhaps *the* defining characteristic – is that the region where life processes go on inside the cell is not in chemical equilibrium with its surroundings. Equilibrium equals death. Life maintains itself in a non-equilibrium state. The American biologist Lynn Margulis sums this up by saying 'life is a self-bounded system.'

This has deep implications for our understanding of the origin of life on Earth. It is now widely accepted that a rain of meteorites and comets delivered the basic components of life to Earth at the end of the early bombardment. Most suggestions about how the transition from non-life to life occurred involve the production of the molecules of life first (DNA, RNA and proteins, though not necessarily in that order), followed by the 'invention' of the cell. One popular idea is that complex compounds were concentrated in a thin layer of material, either trapped in a layer of clay or spread across a surface, and chemistry did the rest. Another is that the crucial chemical processes took place in a hot, chemically rich environment like the ones found today in deep-sea vents, where superhot water is produced by volcanic activity. There are other variations on the theme, all of which hark back to Charles Darwin's speculation, in a letter he wrote to Joseph Hooker in 1871, that:

> We could conceive in some warm little pond, with all sorts of ammonia and phosphoric salts, light, heat, electricity, etc., present that a protein compound was chemically formed, ready to undergo still more complex changes.

But what would happen to such a compound? It would be more likely to be washed away, or destroyed, than to combine with other complex molecules and do something interesting. The interesting things that could lead to life would only take place in a sheltered environment. Where better than inside a cell? To my mind, it is far more likely that cells came first, in the form of bubbles made of semi-permeable membrane, and were taken over by life. And there is strong evidence to support this contention.

Researchers at NASA's Ames Center carried out experiments at the end of the twentieth century in which vacuum-sealed chambers about the size of shoe boxes were cooled to 10 degrees above the absolute zero of temperature, equivalent to −263 °C. A mixture of water, methane, ammonia and carbon dioxide in the chamber was allowed to freeze onto pieces of aluminium or caesium dioxide in the chambers, simulating the way ices form on dust grains in interstellar clouds. Then, the mixture of molecules was bathed in ultraviolet radiation, simulating the radiation from young stars. You will not be surprised to learn that the result was the production of a whole slew of organic molecules, including alcohols, ketones, aldehydes, and very large molecules with as many as forty carbon bonds linking their atoms. The news soon attracted the attention of another researcher, David Deamer, of the University of California, Santa Cruz. Years before, in the 1980s, Deamer had caused a sensation among astrobiologists with his studies of a piece of rock from space called the Murchison meteorite, which landed in Australia in 1969. Searching for signs of organic material, Deamer had ground some of the rock from the meteorite to dust, and washed it over with water to rinse out any organic molecules. To his astonishment, he found hundreds of microscopic globules floating in the water, each one made up of a double 'skin', like little balloons. And when he took some of the frozen material from the Ames experiment and placed it in warm water, Deamer found exactly the same kind of balloons, or bubbles, called vesicles. They were between 10 and 40 micrometres across, about the same size as red blood cells, and essentially indistinguishable from the vesicles obtained from the Murchison meteorite. They were like cells, but without the chemistry of life.

Further research revealed what was going on to make the vesicles. Some of the more complex molecules produced by the action of ultraviolet light on the ice grains, both in the lab (for sure) and in space (presumably), are members of a family known as lipids, with a distinct 'head' and 'tail' structure, like a tiny tadpole. The head end of the molecule is attracted to water, but the tail end is repelled by water.

When such molecules are put in water, they naturally form a double layer, with the head outwards and the tails inwards. And these double-layered 'walls' promptly curl up into tiny balls. This must have happened in the warm waters of the young Earth, trapping things like amino acids and sugars inside the vesicles, in a contained environment where it was possible for the processes that led to life to take place. Without those barriers, the important molecules of life would have been so diluted in the ocean that no interesting chemistry would have taken place. The icing on the cake is that given the raw materials, little balls like this grow, by inserting more lipids into the skin of the bubble, and if they grow big enough they spontaneously divide into two spheres.

It's even possible – though this is not essential to the explanation of how life got started on Earth – that vesicles exist inside comets, so that life itself, not just the precursor chemistry of life, was carried down to Earth at the end of the early bombardment. A team of researchers from NASA's Ames laboratory (Max Bernstein, Scott Sandford and Louis Allamandola) discussed something similar in an article in *Scientific American* in July 1999:

An intriguing possibility is the production, within the comet itself, of [organic material] poised to take part in the life process . . . there are repeated episodes of warming for periodic comets such as Halley when they approach the Sun [and] ample time for a very rich mixture of complex organics to develop . . . It is even conceivable that liquid water might be present for short periods within the bigger comets . . . it is quite plausible that comets played a more important active role in the origin of life.

The most extreme version of this idea was developed in the 1970s and 1980s by Fred Hoyle and Chandra Wickramasinghe, who were convinced that Earth was seeded with life from space. Whether or not these speculations are correct, the important point to emphasize is that the conservative view today is that the Earth was seeded with complex organic molecules, one step from being alive, very early in its history. Life on Earth did not have to get started from scratch from a mixture of simple molecules such as carbon dioxide, water and methane.

It still may not seem likely that the step from non-life to life could occur, even within the protecting enclosure of a vesicle. But there must have been many hundreds of billions of vesicles on the young Earth, and it only had to happen once! As Darwin says, near the end of *Origin of Species*:

> probably all the organic beings which have ever lived on this earth have descended from some one primordial form, into which life was first breathed.

This insight has since been amply confirmed by studies of the genetic material – the DNA – from many different kinds of living thing, and of the workings of the cell itself, which uses exactly the same basic chemistry (for example, to process energy) in all living things. Interestingly, this evidence also suggests that the 'primordial cell' was a heat-loving bacterium that may have come to life near an underwater volcanic vent. Since the young Earth was covered in water and highly active geologically, this is not very surprising and is entirely consistent with the idea that the chemistry of life developed inside a vesicle from space. It just tells us where that vesicle was when life began.

THE COSMIC LOTTERY AND THE DRAKE EQUATION

Can all of this tell us anything about the likelihood of life elsewhere in the Universe? Since we only have the example of the Earth to go by, you might think that it is impossible to draw cosmic conclusions about the existence of life elsewhere. Perhaps it is very easy for life to get started, and all Earth-like planets harbour life; or perhaps it is very difficult for life to get started, and ours is the only inhabited planet. Either possibility, and every possibility in between these extremes, is consistent with the evidence that life exists on one planet – as the statisticians say, you can't generalize from a sample of one. Or perhaps you can. Some astronomers, notably Charles Lineweaver, of the Australian National University, argue that the speed with which life got started on Earth is an important additional piece of evidence to add to the fact that there is life here at all.

Most astronomers agree that the Moon was formed in a huge

impact between a Mars-sized object and the Earth, about 100 million years after the formation of the Solar System. The energy from this impact would have melted the Earth's crust and sterilized the planet, providing a 'blank slate' for the beginning of life on Earth. After that event, for about 600 million years the Earth was subjected to a heavy but decreasing bombardment from space, and it is possible that life got started several times and was wiped out several times before the bombardment ended. There is some controversial evidence that there was a final catastrophe called the 'Late Heavy Bombardment' before this process ended, but Lineweaver does not find any evidence to support this idea, and whether or not the Late Heavy Bombardment occurred the subsequent argument is not affected. Either way, as the long bombardment ended life began. Lineweaver makes the scientific case for the common-sense argument that if life were rare in the Universe it would be unlikely 'that biogenesis would occur as rapidly as it seems to have occurred on Earth'.

He uses the example of a lottery to indicate how the statistics work. If a gambler buys a lottery ticket every day for three days, and loses on the first two days but wins on the third, a statistician would conclude that the odds of winning are unlikely to be close to 1, and are more likely to be about 1 in 3 than about 1 in 100. If a large number of gamblers each buy lottery tickets for 12 consecutive days, and then we pick one of them and discover that he won at least once during the first three days, we conclude that it is likely that the odds of winning are quite good. Statisticians can specify just how good the odds are, in terms of the 'confidence level'. A 95 per cent confidence level, for example, means that their conclusions are likely to be right 19 times out of 20, since 5 per cent is one twentieth of 100 per cent. The 95 per cent level is usually regarded as a good benchmark. In this particular example, the proper statistical conclusion is that the chance of winning the lottery with any one ticket is at least 0.12, at the 95 per cent confidence level. This means you could expect to win slightly more than once in every ten goes. So if the prize was worth at least ten times the cost of a ticket, it would be well worth entering this particular lottery.

As far as biogenesis is concerned, all the Earth-like planets in the

Milky Way are our sample of gamblers, and Earth itself is our chosen individual, who won the lottery very early on. Life emerged (the lottery was won) certainly within 600 million years of the end of the bombardment, and perhaps much sooner than that. The statistical calculation then leads to the conclusion, at the 95 per cent confidence level, that life exists on at least 13 per cent of all the Earth-like planets that are at least a billion years old, and that the proportion of such planets with life is 'most probably close to unity'. Or, in everyday language, life is common in the Universe.

The same argument applies whether life originated on Earth itself, or originated in space and was brought down to Earth in comets, or originated on some other planet and has spread through the Galaxy by hitching a ride on pieces of cosmic rubble (panspermia) or even by being deliberately spread by intelligent aliens (directed panspermia). No matter how life on Earth got started, it got started so quickly that it is highly likely that it exists on other planets as well. This is seen by some people as an important piece of information to feed in to an equation devised by the astronomer Frank Drake to quantify the chance of finding intelligent life elsewhere in the Universe. I have my doubts about the usefulness of this 'Drake equation'; but it does at least provide a neat way of summing up our ignorance about the subject.

The Drake equation was a product of the excitement about space generated by the launch of the first artificial Earth satellites in the late 1950s. At that time, Frank Drake was working at the Green Bank radio observatory, in West Virginia, and was interested not only in the possibility of other intelligent civilizations existing, but in the prospect of communicating with them using radio telescopes. Early on, the handful of astronomers who took the idea seriously coined the term Communication with ExtraTerrestrial Intelligence, or CETI, to describe what they hoped would be the outcome of their work, and in 1974 signals were beamed into space from the giant radio telescope at Arecibo, in Puerto Rico, in the hope that one day some being 'out there' might detect them and reply. 'Some day' will be a long time coming – for reasons I've never understood the signal was beamed towards a cluster of stars known as M13, in the direction of the constellation Hercules. The cluster is 25,000 light years away, and radio waves travel at the speed of light (c), so even if any aliens pick up the message and reply immediately their answer won't get back here until 50,000 years from now! Even that would be too soon for many people. The prospect of actually communicating with ET caused so much alarm in some quarters that the perception of politicians and the public, in the USA in particular, that such beings might not be friendly led to a slight change of name, to the Search for ExtraTerrestrial Intelligence, or SETI. Nowadays, some radio astronomers still listen out for alien signals, but they do not broadcast messages to the stars. It's just too bad if all other civilizations are equally paranoid, so that everybody is listening and nobody is speaking!

That change of name still lay in the future, though, when Drake organized the first scientific conference devoted to the possibility of CETI (or SETI), at Green Bank in 1961. The aim of the meeting was to raise awareness of the possibilities and provide propaganda for fund-raising for the search, and Drake succeeded brilliantly in this aim by coming up with, as the basis for discussion at the meeting, the equation that now bears his name.

This is based on the fact that the probability of two things each happening is equal to the probability of one thing – one event – happening on its own, multiplied by the probability of the other event happening. The chance of rolling a three on a die, for example, is $1/6$, since there are just six possible ways the die can end up. The chance of rolling a three on a second die is also $1/6$. So the chance of rolling two dice and getting two threes is $(1/6) \times (1/6)$, or $1/36$. Simple. And the same basic rule applies if you have a string of events and want to know the probability of all of them happening together.

Drake tried to think of all the factors that could affect the emergence in our Galaxy of a technological civilization capable of communicating across interstellar space. He started with the number of stars in the Milky Way, which he called N^*. Next came the fraction of stars that are like the Sun, f_s. Then, we have to estimate the fraction of those stars that have planets, f_p. The fraction of those planets that lie in the life zone around their parent star is denoted by n_e. Lineweaver's calculation comes in to the next number, the fraction of those planets on which life actually does arise, f_i. Pure guesswork comes in to any estimate of the fraction of those planets where intelligence like ours arises, f_c. And as a kind of joker in the pack, Drake added in a number to provide scope for a guess of the percentage of the lifetime of a planet during which it is occupied by a civilization able and willing to communicate with other intelligences; he called this f_l.

Putting everything together, if N is the number of civilizations beyond Earth that we might be able to communicate with in the Milky Way today,

$$N = N^* \times f_s \times f_p \times n_e \times f_i \times f_c \times f_l$$

This is the Drake equation.

The good news is that we start out with a large number of stars – N^* is at least several hundred billion, and may be a trillion. Even better, although no extrasolar planets were known in 1961, we now know that planetary systems are common, although the jury is still out on whether or not Earth-like planets are common. And if we take Lineweaver's statistical argument at face value, f_i is probably close to 1 and almost certainly bigger than 0.13. The bad news is that if even one of the other numbers in the equation is zero, then N = 0, no matter how big all the other numbers are.

Almost as bad, it isn't immediately obvious how you can quantify those other numbers, although that is more or less what I shall be trying to do in the rest of this book. Worst of all, though, Drake made a gross oversimplification (entirely justified in the context of what he was trying to do and what was known in 1961, but no longer valid) by using just one number to represent an estimate of the fraction of those planets where intelligence like ours arises, f_c. This number is itself best understood as the product of a long string of other numbers representing the probability of the various events in the history of life on Earth that led to the emergence of our own civilization, and, as I shall explain, any one of those numbers could be close to zero, making f_c itself, and therefore N, absolutely tiny.

Human experience suggests that the last number in the equation, f_l, is also small. American Michael Shermer has a take on what civilization is that is different from that of many other people. From one point of view, all of human civilization is a continuous whole; but Shermer points out that many separate civilizations have risen and fallen since records began, and only one of them produced the kind of technology that enables, for example, the construction of radio transmitters. He has looked at the lifetimes of sixty terrestrial civilizations, ranging from Sumeria through Babylonia, Egypt, Greece and Rome and into the modern era, including eleven in China, four in Africa, three in India, two in Japan, six in Central and South America, and six modern states in Europe and America. The total lifetime covered by the sixty civilizations is 25,234 years, giving an average lifetime, f_l, of 420.6 years. Worse, for the twenty-eight civilizations since the fall of Rome the average lifetime is only 304.5 years – and the ability to send and receive radio signals has only arisen once. On that basis, Shermer calculates that even with optimistic estimates for the other numbers in the Drake equation, there are at most three other radio-transmitting civilizations in the Galaxy today.

None of this has stopped people trying to use the Drake equation, or some variation of it, to calculate N, using their own preferred guesses for the numbers on the right hand side of the equation. It's a sign of how futile this approach is that the 'answers' they get range from zero to a few hundred billion. The Drake equation is best regarded not as an equation that can actually be solved to give a realistic measurement of the number of technological civilizations in our Galaxy, but as a kind of mnemonic to remind us of the sort of things we have to take into account when considering the possibility of finding intelligent life elsewhere. Perhaps it isn't surprising that, faced with the complexities involved, some people prefer to go back to arguments based on statistics and probability theory, with some rather startling conclusions.

THE INSPECTION PARADOX AND THE COPERNICAN PRINCIPLE

One of the most crucial numbers in the Drake equation is f_l, corresponding to the lifetime of a technological civilization. Your best guess for the value of this number probably depends on how optimistic you are about the fate of our own civilization. No less an authority than the then President of the Royal Society, Sir Martin Rees, was gloomy enough to suggest that we may have less than a hundred years left; but it is possible to find optimists (many of them science fiction authors) who believe that humankind can overcome all the problems facing us and develop the resources of our Solar System for millions, perhaps even tens of millions, of years. Probability theory can improve on such hunches and give an insight into the likely lifetime of our own civilization. It can also provide insight into our past, using the same statistics that apply to catching a bus.

We've all had the experience of waiting ages for a bus, and then seeing two (or even three) arriving one behind the other. Somehow, most of the times you go to catch a bus you seem to have to wait longer than half of the average interval between buses. How can this be so? Surely it ought to even out, with the next bus coming soon on some days and a longer wait on others? But a little thought shows that it really is true that on average you wait 'longer than average' for your bus.

It happens like this. Suppose the buses on your route start out from the depot at regular 10-minute intervals. They may get bunched up by traffic, but the average interval is still 10 minutes – if one bus is delayed by 2 minutes, the interval between this bus and the one in front is now 12 minutes, but the interval between it and the one behind is now 8 minutes, so the average interval is still 10 minutes. If you stood at the bus stop all day and measured the intervals between each bus, that's what you would find – varying gaps, but with an average of 10 minutes. But that is not what happens when you catch a bus. You arrive at the bus stop at some random time, and get on the first bus that turns up. If the buses were evenly spaced, your average wait would be 5 minutes – half the interval between buses. But because the buses are not evenly spaced, you are more likely to arrive at the stop during a long gap than during a short gap. If there is a gap of 20 minutes after one bus, and then two buses arrive within a minute of each other, it is 20 times more likely that you will arrive at the stop during the long gap than during the short one. So you will probably have to wait longer than half the average interval between buses. Put that way, it's common sense; the probability theorists can quantify all this, and have given the phenomenon the name 'the inspection paradox'.

The inspection paradox can also explain why there were a lot of old people around in Shakespeare's day even though the life expectancy at the time was low. The average life expectancy at birth was low because many children and babies died in infancy. If someone survived to become an adult, they had a good chance of living to a decent old age. Even in a population where the life expectancy at birth is, say, 20, there will be people who live to be 70 or older. No matter how old you are, you will still have some life expectancy, and at any age your total life expectancy is greater than the life expectancy at birth for the general population. Everybody has better than average life expectancy – another example of the inspection paradox. Everybody in good health reading this can take comfort in the knowledge that you have a better than average life expectancy. By the same token, the fact that our civilization is still 'alive' means that it has a better than average life expectancy compared with all the civilizations that have ever existed in the Milky Way Galaxy – but this is based on the idea, which Michael Shermer disagrees with, that all of human history represents a single 'civilization'. As the mathematician Amir Aczel has put it, referring to the longevity of life on Earth, not just of human civilization, but taking this perspective, 'our ability to inspect ourselves is an outcome of the fact that we've been here on this planet for a long time [and] the conditional probability that we have been around for longer than other civilizations . . . is high.' And 'assuming other civilizations exist, chances are that we are among the first in our galaxy to arrive at this level of advancement.' Curiously, he has arrived at the same conclusion as Shermer – that we are probably alone in the Galaxy – starting out from a completely different assumption about what constitutes a civilization!

But that only tells us about the past. What about the future of human civilization?

In 1993, Richard Gott, of Princeton University, provoked debate when he published a paper in the leading science journal *Nature* in which he used this kind of statistical reasoning to estimate the total lifetime of our species. He found that at the 95 per cent confidence level *Homo sapiens* is likely to be around for a total (including our history to date) of between 200,000 years and 8 million years; the lower end of this range corresponds to us going extinct more or less tomorrow, and even the upper end is hardly impressive on an astronomical timescale.

The only assumption that Gott put in to his calculation is that you are a random example of all the intelligent observers that ever exist. He calls this the 'Copernican principle', by which he means the assumption that we do not occupy a special place in the Universe – in fact, Copernicus only said that we do not occupy a central place in the Universe, but the extension of this idea to say that we occupy a non-special place is quite common. Slightly less grandiosely, it is sometimes called the 'Principle of Terrestrial Mediocrity' – the idea that we occupy an ordinary planet orbiting an ordinary star in an ordinary galaxy.

Before setting out the consequences of this assumption, Gott illustrates the power of this approach with two examples from his own life. The essence of the argument is that if you observe something at random for the first time, there is a 95 per cent chance that you are seeing it in the middle 95 per cent of its lifetime. Just 2.5 per cent of the time you might be seeing it near the beginning of its life, and just 2.5 per cent of the time you might be seeing it near the end of its life. If you know how old the thing is when you first see it, you can use these figures to set limits on its future lifetime – with standard probability calculations similar to those involved in the bus 'paradox', the future lifetime of the thing you are observing should lie between $1/39$ times and 39 times its past lifetime, at the 95 per cent confidence level.

In 1969, on a trip to Europe, Gott saw both Stonehenge and the Berlin Wall for the first time. Stonehenge is about 3,700 years old, and in 1969 the Berlin Wall was 8 years old. This kind of calculation would imply a future lifetime for Stonehenge of at least a hundred years, and a future lifetime for the Berlin Wall (as of 1969) of no more than a couple of hundred years. The Wall fell in 1989, but Stonehenge is still there, in line with these estimates.

Applying the same logic to humankind, Gott starts from the estimate, based on fossil evidence, that modern people, *Homo sapiens sapiens*, have been around for 200,000 years. This suggests that we are likely to be around for at least another 5,000 years, but for no more than 8 million years. Gott points out that the average lifetime for a mammalian species is about 2 million years, and that our immediate ancestor, *Homo erectus*, was around for just under 1.5 million years, so his calculation is very much in line with what we know from the fossil record. Perhaps our descendants will evolve into something as different from us as we are from *erectus*, and still have a technological civilization; but further calculations by Gott paint a much more gloomy picture.

The same kind of statistical arguments apply to the number of people alive on Earth today, compared with the number already born and the number yet to be born. In this case, you are a random 'observer' picked out from the population of all the people that ever were or ever will be simply by being born. The probability equations tell us that 50 per cent of all such 'observers' are born at a time when the population is at least half of its maximum value. The present human population of the Earth is nearly 7 billion, and the planet's estimated carrying capacity is about 12 billion, so on that basis it is not surprising that you are alive today. It is an example of the fact that you are a random intelligent observer picked out by chance from all of the intelligent observers in the past, present and future. In 1993, using a version of the calculation previously applied to timescales, Gott calculated that the number of people still to be born was at least 1.8 billion, and estimated that this total would be reached in the first decade of the twentieth century. Applying the same calculation today, at least another 2 billion people are still to be born, and this will take roughly another ten years. If Gott is right, sooner or later (and most probably sooner) there will be a population crash and a collapse of civilization. That may not apply with such force to other civilizations; but Gott has more bad news for proponents of SETI.

SETI advocates still pin their hopes mainly on radio communication. In 2004, one of the most prominent of these advocates, the Microsoft billionaire Paul Allen, donated a further $13.5 million, on top of earlier donations totalling $11.5 million, towards the construction of the

'Allen Array', a dedicated SETI radio telescope. But he may have been wasting his money (although, fortunately, the Allen Array can be used for conventional radio astronomy as well). Our civilization has only used radio for about 120 years, and the probability calculation says that our future lifetime as a radio-transmitting civilization is likely to be between three years and five thousand years. This doesn't necessarily imply the collapse of civilization – our descendants may move on to something superior to radio communication, just as we no longer use smoke signals. But it does suggest a severe constraint on our chances of making contact with ET.

Assuming that we are a typical example of a radio-transmitting civilization, and plugging this number into his version of the Drake equation, Gott estimates that the number of radio-transmitting civilizations in the Galaxy today is no more than 121. The chance of one of them being within range is so small as to make any radio SETI project futile, if Gott's calculations are correct. It may be futile in any case, since the most likely flaw in his argument is the Copernican principle itself. Perhaps we are not typical observers, after all. That's the most likely resolution of the most famous ET 'paradox' of them all, formulated clearly by the physicist Enrico Fermi, and now given his name, although he was not, in fact, the first person to puzzle over it.

PANSPERMIA AND THE FERMI PARADOX

Fermi was one of the most important physicists of the twentieth century. Among his many achievements he predicted the existence of the particle known as the neutrino, and he received the Nobel Prize in 1938 for his work on radioactivity and nuclear reactions. With war looming in Europe, instead of going back to Fascist Italy, the Fermi family went on from the Nobel awards ceremony in Stockholm to the United States, where Fermi became the leader of the group at the University of Chicago that built the first nuclear reactor, known at the time as an 'atomic pile'. The pile 'went critical' at 2.20 p.m. on 2 December 1942.

Fermi had a great ability to see to the heart of a problem, and to express complex ideas in simple language. He was a master of the art of making rough estimates – called order of magnitude calculations – of the solution to complicated problems, and it was this that led him to the Fermi paradox. He became so well known for this that these kind of puzzles are often referred to as 'Fermi questions'. A simple example of a Fermi question is: How many barley sugar sticks will fit into a 1-litre jar? The point is not to get a precise answer, but to make an educated guess. A stick is roughly cylindrical, about 2 cm long and 1.5 cm in diameter (0.75 cm in radius). The volume of such a cylinder is π multiplied by the square of the radius multiplied by the length, which comes out as about $25/7$ cubic cm, if we use the approximation $\pi = 22/7$. But the sticks don't fit tightly together in the jar, so as a guess we might say that 20 per cent of the volume is air. A litre is 1,000 cubic cm, so the sticks actually take up 800 cubic cm, and the number of sticks needed to do the job is 800 divided by $25/7$, which is about 220 (it's exactly 224, but it would be silly to quote the 'answer' that accurately). It's this kind of calculation that astronomers use to estimate things like the number of stars in a galaxy without actually counting them all, and which Fermi used to come up with his famous puzzle.

Although Fermi died in 1954 at the age of only 53, without leaving a memoir of the occasion when he came up with his paradox, the exact details of the story were reported by the physicist Eric Jones, on the basis of interviews with Fermi's contemporaries, in the August 1985 edition of the journal *Physics Today*. It happened in the summer of 1950, when Fermi was at the Los Alamos laboratory where the first nuclear bomb had been developed a few years earlier. This was during the height of public interest in flying saucers (UFOs), triggered by sightings of secret aircraft during the post-war period. There had also been a spate of disappearances of rubbish bins (trash cans) from the streets of New York, and the *New Yorker* had just published a cartoon suggesting that the trash cans were being stolen by aliens. Fermi and his colleagues had been laughing over the cartoon on the way to lunch, and this led them into a discussion about the (im)possibility of travelling faster than light. Over lunch, the conversation turned to other matters. Then, suddenly, Fermi asked out loud, 'Where is everybody?' His colleagues realized that he was referring to extraterrestrial intelligences, and since it was Fermi asking the question, they took it seriously. He quickly made an order of magnitude calculation which implied that even if they were restricted to travelling slower than the speed of light, aliens should have long since colonized the entire Galaxy and the Earth should have been visited many times. The neatest summing up of the Fermi paradox is contained in the question, 'If they are there, why aren't they here?' In other words, if there are extraterrestrial intelligences, why haven't they visited us?

Nobody took much notice of the puzzle, except as a topic for coffee-time discussion, until 1975. Then, like two buses coming along together, two scientific papers appeared that stimulated a much broader discussion. Writing in the *Journal of the British Interplanetary Society*, David Viewing re-stated the puzzle, giving full credit to Fermi. The same year, Michael Hart published a paper in the *Quarterly Journal of the Royal Astronomical Society* in which he phrased the question slightly differently – why are there no intelligent visitors from other worlds on Earth today? Unlike Viewing, though, Hart offered four possible categories of explanation for the puzzle:

1. It may be physically impossible to get from there to here
2. They are there, but they have no wish to contact us
3. They are there, but they have not yet had time to reach us
4. They have been here but have left no trace and are not here now

There's one other possibility, which is in essence a variation on Hart's category 4 – perhaps we are the aliens. This is another way of looking at the idea of panspermia, mentioned earlier. And although it does not affect the main argument of this book, it is not only fascinating in its own right, but offers a powerful insight into how easy it is for life to spread across the Milky Way, given the enormous span of cosmic time available.

Panspermia literally means 'life everywhere', and speculation about the possibility of life everywhere goes back to ancient times. But the science of panspermia can be seen to have started with some remarks made by William Thomson, later Lord Kelvin, in his Presidential

Address to the meeting of the British Association for the Advancement of Science, in 1871. Thomson was following up work by Louis Pasteur in the 1860s, which had finally proved that living things do not spontaneously come forth from non-living things – that maggots, for example, are not a product of rotting meat, but come from eggs laid by flies. He declared that all life comes from life – all living things have ancestors. As Thomson put it, 'dead matter cannot become living without coming under the influence of matter previously alive. This seems to me as sure a teaching of science as the law of gravitation.' We would now say that there was, at least once and very long ago, an occasion when a living molecule arose from non-life, but that does not affect the subsequent thrust of Thomson's argument.

Thomson made an analogy with the way life appears on a newly formed volcanic island. 'We do not hesitate to assume that seed has been wafted to it through the air, or floated to it on a raft,' rather than being generated spontaneously from the dead rocks. The Earth, he said, is in the same situation:

> Because we all confidently believe that there are at present, and have been from time immemorial, many worlds of life besides our own, we must regard it as probable in the highest degree that there are countless seed-bearing meteoritic stones moving about through space. If at the present instant no life existed upon earth, one such stone falling upon it might, by what we blindly call natural causes, lead to its becoming covered with vegetation.

Few of Thomson's contemporaries took the idea seriously. But one who did was the Swede Svante Arrhenius, a chemist who won the Nobel Prize in 1903 for his work on electrolysis, and who was a pioneering investigator of the atmospheric 'greenhouse effect.' His grandson, Gustaf Arrhenius, was, incidentally, one of the people who studied the isotopic evidence for early life on Earth.

Instead of considering seeds being carried through space inside meteorites, Svante Arrhenius speculated that microorganisms like bacteria might be carried high into the Earth's atmosphere and escape, to be blown across space by the pressure of the Sun's radiation. Such microorganisms can remain inert for long periods of time before reviving, perhaps long enough for them to cross interstellar space and land on some other Earth-like planet. Arrhenius estimated the travel times for such seeds of life, starting from Earth, as 20 days to Mars, 80 days to Jupiter, and 9,000 years to Alpha Centauri, the nearest star to the Sun. And if they could travel one way, why not the other, with the Earth having been seeded by microbial life from space?

Panspermia has never been a mainstream idea in astrobiology, but since the speculations of Thomson and Arrhenius people have returned to it from time to time in different contexts, and strengthened its scientific credentials without ever making a completely convincing case. One variation on the theme picks up Arrhenius' original idea, and adapts it to take account of the conditions that we now know life would be likely to encounter on its journey through space. A star like the Sun produces a large amount of ultraviolet radiation, which would be lethal for microorganisms that escape into space. But Jeff Secker, of Washington State University, and his colleagues Paul Wesson and James Lepock, at the University of Waterloo, Canada, calculated what would happen if the microorganisms were encased in tiny grains of ice or dust. This would still not be enough to protect them from the radiation from a star like the Sun. But in old age a star swells up to become a so-called red giant, which would be luminous enough to push the particles out across space, but does not produce the damaging ultraviolet rays. Secker and his colleagues calculate that the journey time for such grains carrying the seeds of life across space is about 20 light years per million years. Since there are dozens of stars within 20 light years of the Sun, this suggests that life could move easily from one planetary system to the next, and could cross the entire Galaxy, spreading life everywhere, in a few billion years.

The better the protection, of course, the longer the organisms could survive. Jay Melosh, of the University of Arizona, has shown that microorganisms could survive for many millions of years deep inside large chunks of rock. This is intriguing within the context of our own Solar System, because large impacts from space can eject just such chunks of rock from the surface of a planet. Indeed, there are meteorites found on Earth which have been identified, partly from isotope evidence, as coming from Mars. One of these pieces of Martian rock, dubbed ALH 84001, has been the subject of intense investigation since claims were made that microscopic tube-like structures in the rock resemble fossilized bacteria. There is very little evidence to support this claim. But what the presence of meteorites like ALH 84001 on Earth does tell us is that if there were life on Mars it could be carried to Earth in this way, after spending millions of years wandering about the inner Solar System after the impact in which it was blasted off the surface of Mars. It's slightly harder for life from Earth to get transported to Mars, or anywhere else, in this way, because of the Earth's stronger gravitational pull. And, alas, it is virtually impossible for large chunks of rock to be completely ejected from the Solar System, to carry the seeds of life to other planetary systems.

It would be a lot easier if the seeds of life were deliberately directed at suitable planets. This idea, called directed panspermia, was developed by Francis Crick, co-discoverer of the structure of DNA, and another molecular biologist, Leslie Orgel. It would certainly be easy, using technology only slightly more advanced than our own, to send out probes packed with the kind of blue-green algae that have proved so successful as life forms on Earth towards any interesting looking planetary systems. And it would be a quick way to 'colonize' the Galaxy, if you feel that blue-green algae are suitable representatives of life on Earth. So are we the colonists? Did our primordial ancestral cells arrive on Earth as a gift from another civilization? Are 'they' here because we are them? It's certainly technically possible. But the big question is why would any intelligent beings do this? Even Crick and Orgel have never been able to give a satisfactory answer to that one, and in his book *Life Itself* Crick admits that 'as a theory it [directed panspermia] is premature.' Orgel is equally undogmatic. He has been quoted as saying 'my opinion is that we have no way of knowing anything about the possibility of life in the Cosmos. It could be everywhere, or we could be alone.'

I agree with Crick's conclusion. But I have not ignored the idea of directed panspermia, because it introduces the notion that, for whatever motives, an alien civilization, or civilizations, might decide to send spaceprobes to visit other planetary systems. This possibility provides what I consider to be the definitive resolution of the Fermi paradox. Before unveiling that resolution, however, it is only fair to mention the other suggested 'answers' to the puzzle, if only to say why they are implausible.

The most detailed and accessible compilation of proposed 'answers' to the Fermi paradox has been made by Stephen Webb, a physicist at the Open University, in his book *Where is Everybody?* If the few examples I have room for here whet your appetite for such speculation, that is the place to look for more, although the weirdness of some of the suggestions only adds strength to my own argument.

Many people, of course, believe that aliens are here, and that some of them are even in the habit of abducting people in their flying saucers. There is no credible evidence for this, and although in some quarters the term Unidentified Flying Object, or UFO, has become synonymous with 'flying saucer', UFO buffs are missing the point that just because something is unidentified doesn't automatically mean that it is an alien spacecraft. If a car goes past me too fast for me to identify the make, in a sense it is an unidentified moving object; but that doesn't automatically mean that it is a top-secret rocket-propelled prototype. The logical suggestion is that on closer inspection I would be able to identify it as a known make of car. So many reported UFOs turn out on inspection to be identifiable, as known atmospheric phenomena, weather balloons, aircraft, and the like – or even (I kid you not) the planet Venus – that the simplest explanation for the small percentage that remain formally unidentified is that they are also caused by natural phenomena, but we don't have enough information to decide which category they belong to, just as I don't have enough information to decide the make of the fast car.

A closely related idea to the flying saucer story is the suggestion that we live in a cosmic zoo, or a kind of wilderness area where primitive creatures such as ourselves are being left to develop at their own pace. This often carries with it the implication, explicit in many science fiction stories including the *Star Trek* movie franchise, that once we reach a certain level of technological ability (or perhaps, once we grow up enough to stop fighting among ourselves) we will be welcomed into some sort of galactic club of advanced civilizations, albeit as a junior member. Among the many objections to this idea is: why didn't the galactic club take over the Earth in the billions of years when it was only occupied by single-celled life forms? Did they only discover us a few million years ago, when primates were clearly already on the road to intelligence? And why is there absolutely no evidence of their activity out among the stars of the Milky Way?

At the other extreme from the galactic club there is the suggestion that all the alien civilizations have stayed at home, because they are not interested in space travel (a variation on the theme, which I alluded to tongue-in-cheek earlier, that everyone is listening for a signal but nobody is transmitting). This argument might hold water if we are thinking about just one alien civilization. But if you want to believe that what we call civilization is common in the Galaxy, you would then have to accept that none of them want to investigate their surroundings, because it turns out, as I shall explain, to be very easy to make a mark on the Milky Way. We will soon do so ourselves, if we give the lie to Martin Rees's gloomy prognostications and survive the present century.

The most serious objection to the idea that 'if aliens are there, they ought to be here' is that interstellar travel is tedious and difficult – even though the whole point of Fermi's insight is that he realized, from simple order of magnitude calculations, that on the timescale of human civilization (let alone the timescale of life on Earth) it is relatively quick and at one level easy. There are even examples from human history – strictly speaking, prehistory – that highlight the point.

What's difficult about travelling to the stars is doing it within a human lifetime, and it's more than doubly difficult to do so and come home again. We should not ignore the possibility that other life forms may be much more long-lived than us, and would find a journey lasting a few hundred years no more tedious than we find a flight across the Atlantic, nor the possibility of a civilization so much more advanced than ours that it can take advantage of the shortcuts through spacetime allowed by the general theory of relativity, or tap in to the quantum field energy of empty space to power their ships. Either possibility only makes it even more likely that aliens will colonize the Galaxy. But it could be done by beings like us with only slightly more advanced technology than our own. Possible propulsion systems include nuclear-electric rockets, fusion rockets, the interstellar ramjet, and (my favourite) 'starsailing' with the aid of powerful planet-based lasers. A neat and only slightly dated overview of the possibilities is provided by Robert Forward and Joel Davies in their book *Mirror Matter* (Forward, himself a physicist, is one of the leading proponents of starsailing), and I shall not elaborate on them here. What matters is that whatever your means of propulsion, if you go slowly enough you can get to the nearest stars without too much effort. And if colonies are established on planets orbiting those stars, they can reach the stars near to them but farther from us without too much effort.

The key to colonizing the Galaxy is that you don't come back. The archetypal example from human prehistory is the way Polynesian people spread from island to island across the Pacific, eventually reaching all the way to New Zealand, and even to isolated Easter Island, some 1,800 km from the nearest land. They didn't visit new islands and then go back; they settled on new islands, and used those islands in turn as bases from which to send out people who settled on more remote islands. None of this was planned. It wasn't part of a grand scheme to colonize the Pacific. It just happened, because of population pressure, or from a basic urge to find out what lay beyond the horizon.

In a similar way, although our ancestors evolved in East Africa, their descendants spread out on foot all over the world. They even crossed the land bridge between what are now Siberia and Alaska, and got all the way down into South America. Like the Polynesian voyages, none of this was planned. It just happened that in each generation (or just in some generations) a few people moved on a little way from their neighbours, in search of food and water, or just to see what lay over the next hill, or to get away from the crowd. The whole process took about 30,000 years. Australian-born radio astronomer Ronald Bracewell, from Stanford University, summed this up by pointing out that it was quicker for humans to walk from Africa to South America than it would have been for human-level intelligence to have evolved independently in South America. The last stage of the journey, from present-day Canada to Patagonia, covers about 13,000 km, a distance that could have been traversed in only a thousand years at a rate of merely 13 km per year. His conclusion is that it is much more likely that intelligent life would spread across the Galaxy from the first 'intelligent planet' than that it would evolve independently several (or many) times on several (or many) different planets.

Bracewell's conclusion draws on a calculation made by Michael Hart and published in his paper in the *Quarterly Journal of the Royal Astronomical Society* in 1975. Hart used the example of the possible future colonization of the Galaxy from Earth. Making reasonable assumptions about possible propulsion systems allowed by the laws of physics, he assumes that 'we eventually send expeditions to all of the 100 nearest stars', which all lie within about 20 light years of the Sun. 'Each of these colonies has the potential of eventually sending out their own expeditions, and their colonies in turn can colonize, and so forth.' With no pause between trips, 'the frontier of space exploration would then lie roughly on the surface of a sphere whose radius was increasing at a speed of 0.10 c. At that rate, most of our Galaxy would be traversed within 650,000 years.' Of course, the assumption of no pause between trips is over-optimistic, but even if the length of time between voyages is the same order of magnitude as the length of a voyage, the time needed to colonize the Galaxy would only be doubled, to 1.3 million years. This is not much more than one eight-thousandth of the age of the Galaxy. More pessimistic estimates of the time required, based on how long it takes for colonies to become established, range from a few million years to 500 million years – but even that is still very short compared with the age of the Galaxy. And, as Fermi appreciated, the precise numbers don't matter, just the order of magnitude. If we could do it, so could they. So – why aren't they here?

The argument strikes with even more force once you realize that it isn't even necessary to subject living creatures to the boredom and hazards of interstellar travel.

PROBING FOR AN ANSWER

The most powerful argument for the non-existence of other technological civilizations in our Galaxy stems from the work of two mathematical geniuses who each made major contributions both to the science of computing and to the Allied victory in World War Two. Alan Turing is remembered as a cryptographer who was the leading member of the team at Bletchley Park, in Buckinghamshire, which cracked the German codes during that war. But as early as 1936 he had written a scientific paper, 'On Computable Numbers', which laid out the fundamental principles of machine computing. Turing proved that it is possible in principle to build a machine, now referred to as a universal Turing machine, which could solve any problem that could be expressed in the appropriate machine language. To a generation brought up on personal computers, this seems blindingly obvious – but it was the proof that such machines could be built that started science and technology down the road to the modern computer. In Turing's own words, the universal machine 'can be made to do the work of any special-purpose machine, that is to say carry out any piece of computing' if fed with the right program. The computer built at Bletchley Park as part of the code-breaking effort was an example of a special-purpose machine, designed to do only one job. But the first general purpose computer was soon developed, with the aid of Hungarian-born John von Neumann, who among many other things worked on the Manhattan Project, which developed the first nuclear bomb.

Von Neumann's interest in the idea of a computer that could solve any problem led him to think about the nature of intelligence and of life. Intelligent beings are in a sense universal Turing machines, since they can solve many different kinds of problems, but they have the additional ability to reproduce. Was it possible in principle for there to exist self-reproducing Turing machines in a non-biological sense? Von Neumann proved that it is indeed possible. The process involves just a few simple steps. First, the computer program stored in the memory banks of the machine instructs the machine to make a copy of the program and store it in some sort of memory bank (nowadays, we could imagine this to be an external hard drive). Then, the program instructs the machine to make a copy of itself, with a blank memory. Finally it tells the machine to move the copy of the program from the storage device into the new machine. Von Neumann showed, as long ago as 1948, that living cells must follow exactly the same steps when they reproduce, and we now understand this in terms of nucleic acids as the 'program' and proteins as the 'machinery' of the cell. First, the DNA is copied. Then, as the cell divides in two the copy of the DNA is moved into the new cell.

A self-reproducing non-biological automaton is now often referred to as a 'von Neumann machine'. Neither Turing nor von Neumann lived to see these ideas developed. Turing was only 41 when he killed himself, in 1954, after years of harassment by the authorities for his homosexuality; von Neumann died of cancer in 1957 at the age of 53. It was Ronald Bracewell who suggested that probes could be used to explore the Milky Way, and the American Frank Tipler who presented the full force of the argument which shows how quickly such von Neumann machines could visit every interesting planet in the Galaxy.

The key point is that a technological civilization only has to build one or two probes in order to colonize (by proxy) the entire Milky Way. Such a probe would be programmed to use the raw materials that it found among the asteroids and other cosmic rubble in a planetary system, plus the energy of the parent star, to build copies of itself, and send those copies off to explore other planetary systems. One or two probes sent from Earth to the Asteroid Belt between Mars and Jupiter could mine the raw materials there to make a fleet of identical probes which could set off to explore nearby stars, maintaining contact with home by radio. Each time one of them arrived in a new planetary system, as well as reporting its findings it would set about building copies of itself and repeating the process. Making the very modest assumption that probes could travel at one fortieth of the speed of light and that they were programmed to seek out stars with planets, it would take less than 10 million years (in a galaxy 10 billion years old) from the construction of the first probe to visit every interesting planet in the Galaxy. And all it costs is the construction of one initial probe (or at most a few copies, as backups).

Even with our present technology, we could get a decent-sized probe to the nearest star. It would involve sending the probe on a close flyby of Jupiter, using the gravity of Jupiter like a slingshot to speed it up and send it diving past the Sun, where the Sun's gravity would give it a further boost, sending the probe out of the Solar System at a speed of about 0.02 per cent of the speed of light. When the probe arrived at its target star, it could use the gravity of the star and its planets to slow itself down. It would take several thousand years for the probe to reach its target, but if civilization still existed on its home planet and anyone was interested, it could be programmed by radio to construct not just a replica (or replicas) of itself but an improved, faster version (assuming technology back home had advanced in the millennia since the probe was launched). It might take thousands of years for any interesting news to come back from the first probe, or probes; but as they reproduced and spread faster and faster through the Galaxy news about different planetary systems would come flooding in several times each year.

This is so nearly within our present technological ability that it is quite clear that within a couple of decades at most (unless civilization collapses) we will be able to start this process. And by 'we', I don't necessarily mean the full might of government-sponsored agencies such as NASA. Individuals such as Paul Allen already pay for radio telescopes to search for ET; the Paul Allens of the next generation may well be able to pay to explore (or at least start the exploration of) every planet in the Galaxy. Such an individual might well be hoping to get news from the nearest planetary systems within their own lifetime, rather than caring much about what happens millions of years from now. But since it literally costs no more to explore the entire Galaxy than to explore the nearest planetary system, who could resist going the whole hog?

That is why the possibility of constructing von Neumann probes is such a powerful argument that we are alone in the Milky Way. All the arguments that 'they' are out there but are, for whatever reason, avoiding contact with us require that every technological civilization is working together to keep their presence secret. But it isn't just the case that you only need one civilization to break ranks and send a few probes out into the Galaxy. All it needs is for one individual being to send one probe and every interesting planet will be visited in a few million years.

Fermi asked, 'If they are there, why aren't they here?' The solution to the puzzle is that they are not there. But that raises an even more important question. Not 'Are we alone?' but '*Why* are we alone?' If they are not there, why are we here? What is it that is special about our location in the Universe, in both space and time, that has allowed the development of the only technological civilization in the Galaxy? Why is Earth the only intelligent planet? That is the theme of the rest of this book – the reason why we are here to ask such questions.

2

What's So Special about Our Place in the Milky Way?

Why does intelligent life exist in the Milky Way Galaxy? Our presence is intimately related to the structure of the Milky Way, and to the Sun's location within the Milky Way. Although I am not concerned here with the possible existence of life in other galaxies, understanding the galaxy in which we live, the Milky Way, depends on understanding the way galaxies in general form, and how they change (evolve) as time passes. Astronomers have a good understanding of at least the outlines of all this, because of the way telescopes act as time machines. If an object is, say, a hundred light years away, that means that light takes a hundred years to travel from that object to us, so that we see it today as it was a hundred years ago, when the light set out on its journey. This is known as the 'look back time'. Some astronomical telescopes are so sensitive that they can see galaxies so far away that the light left them just after the Big Bang in which the Universe as we know it was born, about 13 billion years ago. So we can actually see young galaxies soon after they formed. And, of course, we can see galaxies in later stages of their evolution at all lesser distances. We cannot see individual galaxies changing as time passes, but we can see galaxies in all stages of their development, and use this information, combined with our understanding of the laws of physics and computer simulations, to work out how a galaxy like the Milky Way got to be the way it is today. In much the same way, biologists do not have to watch a single oak tree grow from an acorn to understand how mature oak trees get to be the way they are; they can see oak trees in all stages of their development in a forest, and work out the life cycle of an oak tree from these observations.

Dark matter, revealed today by its gravitational influence on bright stars and galaxies, played a key role in the formation of galaxies when the Universe was young. This is a form of matter unlike the atoms and molecules we are made of, revealed only by its gravitational influence. Investigating the nature of dark matter is a major concern of astronomers today, but in the present context all that matters is that it is there, and that without it galaxies like the Milky Way could not have formed. Clumps of dark matter exerted a gravitational pull which attracted streams of atomic matter, like water flowing into a series of potholes in an unmade road. These streams of atomic matter consisted only of hydrogen and helium, produced in the Big Bang, with none of the heavier elements so vital to our existence. The result was huge clouds of gas, beginning to collapse under their own weight, not just because of the influence of dark matter, and getting hotter as they did so, in much the same way that the air in a bicycle pump gets hotter when it is squeezed. The clouds have to stop collapsing when the inward pull of gravity is balanced by the pressure of the hot gas, and for clouds made only of a mixture of hydrogen and helium this happens when the clouds are still very large, because is hard for them to radiate energy away into space. When stars form today, they do so from much smaller clouds, because the presence of heavier elements, such as carbon, allows heat to be radiated away more efficiently and the clouds to shrink. But simulations show that the first stars that formed after the Big Bang had masses hundreds of times the mass of the Sun and surface temperatures of about 10,000 K, producing a glow of radiation that can still be detected today using infrared telescopes flown in space.

A star keeps shining because of nuclear reactions going on in its core, which convert light elements into heavier elements and release energy in the process. The heavier a star is, the more furiously it has to burn its nuclear 'fuel' in this way to hold itself up, so heavy stars are short-lived. Within about 250 million years of the Big Bang, the first stars would have used up all their fuel and exploded, scattering the outer layers of their material, including some of those heavier elements, through the nearby gas clouds. Blast waves from the explosions would have triggered more clouds to collapse, but now the collapsing clouds could shrink more than their predecessors, because of the presence of those heavy elements which allowed more radiation to escape. As they shrank, the clouds broke up into smaller fragments. They made the first stars similar to those we see in the Milky Way Galaxy today. Indeed, some of the stars we see in our Galaxy may be left over from this early stage of galaxy formation. The oldest stars in our Galaxy, containing only a smattering of heavy elements, are estimated to have ages greater than 13.2 billion years, which means that they formed within 500 million years of the Big Bang itself. Stars like these, which contain only a small proportion of heavy elements (lacking in 'metals', in the astronomical sense of the term) are known, for historical reasons, as Population II.

But the environment in which Population II stars formed was already very different from the environment in which the first stars formed, quite apart from the presence of the first metals. When a star bigger than about 250 times the mass of the Sun dies, although the outer layers are blown apart in a great explosion, most of the material collapses to make a black hole with more than a hundred times the mass of the Sun. Since those primordial stars must have formed in the densest concentrations of matter in the Universe at that time, many of these black holes would have been close together, and merged to make even more massive objects. There is a wealth of evidence that all galaxies like our own, including the Milky Way itself, have supermassive black holes at their hearts. It is impossible to say for certain where these black holes came from, but it is most likely that they formed from black hole mergers following the death of the first stars a few hundred million years after the Big Bang.

MAKING GALAXIES

It isn't easy to see any except the brightest astronomical objects at distances corresponding to a look back time of around 13 billion years, but observations of such bright objects (known as quasars) show that black holes with at least a billion times as much mass as our Sun, surrounded by clouds of atomic material in which stars could form, existed before the Universe was a billion years old. There must have been many similar but smaller such objects as well, too faint to see at such vast distances; simulations show how black holes formed the cores on which galaxies grew as clouds of atomic matter were caught in the gravitational grip of the black holes. Observations made with the Hubble Space Telescope have picked out some of these early galaxies, smaller than the Milky Way and very blue in colour because of the presence of many hot, young stars in them. But don't imagine that black holes came first and galaxies grew around them later; the process is better envisaged as a kind of co-evolution, in which the black holes and surrounding galaxies grew together from a single cloud of primordial material. Both the mass of the black hole and the mass of the surrounding galaxy depend on how much stuff there was in the cloud to start with.

The calculations show that this process can produce an object as big as the Milky Way within a few billion years, provided that the central black hole has a mass at least a million times the mass of our Sun. Observations of the central region of the Milky Way show that stars at the heart of our Galaxy are orbiting very rapidly around some unseen central object. The speed with which they are moving tells us that the mass of this central object which has them in its grip is about 3 million times the mass of our Sun, but confined within a radius of about 7.7 million km, only about twenty times the distance from the Earth to the Moon. It can only be a black hole. Everything fits together beautifully. But as far as life is concerned, although the presence of the central black hole was essential for the birth of the Galaxy, what mattered once the Galaxy did form was what went on in its outer regions, far away from the black hole.

All of this explains how the bulge of our Galaxy (the bit like the yolk of the fried egg) formed, and how similar bulges formed in other galaxies. It also explains the origin of smaller members of a family of galaxies known as ellipticals, which are like the bulge of a disc galaxy without the disc (or like the yolk of the fried egg without the white). All primordial galaxies, except for small, irregularly shaped collections of stars unimaginatively called irregulars, seem to have developed as elliptical bulges, but not all of them developed discs.

Presumably, this depended on whether or not there was enough material nearby to be captured into a disc around the central bulge.

This didn't happen all at once. Disc galaxies like the Milky Way seem to have been built up by adding different bits and pieces to the central bulge as time passed. There is direct evidence for this in the way stars move around the disc of our Galaxy. Astronomers have found that although most of the stars in the disc are moving together in a regular way, like a crowd of runners going round a track, they can pick out several long, thin streams of stars which are similar to one another but have slightly different composition from the background stars, and which are moving in the same direction as each other but at an angle to the motion of most of the stars in the disc. There are about a dozen of these streams now known. The amount of material in a stream ranges from a few thousand solar masses to a hundred million times the mass of our Sun, and their lengths range from 20,000 light years to a million light years. They are the remains of small galaxies that came too close to the Milky Way and got disrupted by the gravitational pull of our Galaxy. As they disperse these stars will merge in with the rest of the Milky Way, and become indistinguishable from their neighbours. The implication is that the Milky Way reached its present size through a kind of cosmic cannibalism, swallowing up its smaller neighbours; studies of nearby disc galaxies show exactly the same process going on there as well. Among other things, this explains why there are stars more than 13 billion years old in a galaxy that was born only about 10 billion years ago – it swallowed up those older stars as it grew.

The Milky Way, and galaxies like the Milky Way, were already in existence 3 or 4 billion years after the Big Bang, roughly 10 billion years ago. Although they were still growing as a result of this cosmic cannibalism (and still are!), the distinct bulge and disc structure already existed, with increasingly metal-rich stars, known as Population I, forming in the disc. As time passed, the gas in the disc became increasingly enriched with heavy elements by succeeding generations of stars, so that by the time the Sun formed, about 5 billion years ago, when the Milky Way was half its present age, there were enough 'metals' to make planets and, on at least one planet, people.

This is not the end of the story. Giant elliptical galaxies, many times larger than the Milky Way, seem to have formed from mergers between disc galaxies, in which the structure of the discs has been destroyed. This may well be the fate of our own Galaxy, which is on a collision course with its near neighbour, a galaxy known as M31, in the direction of the constellation Andromeda. But this won't happen for several billion years, and has no direct bearing on the puzzle of why we are here. What is relevant is the story of how stars converted hydrogen and helium into heavier elements and scattered them through space to be included in the raw material of future generations of stars and planetary systems.

MAKING METALS

Even in a star like the Sun, born nearly 10 billion years after the Big Bang, there is only a smattering of heavy elements. In terms of its mass, the Sun is made up of about 71 per cent hydrogen, just over 27 per cent helium, and just under 2 per cent of everything else put together. The inner structure of the Sun can be studied by analysing waves on its surface, in the same way that the inner structure of the Earth can be studied by analysing earthquake waves. Astronomers can also calculate what conditions must be like in the heart of the Sun from the laws of physics, the known size of the Sun and its brightness. Combining these two approaches tells us that half of the Sun's mass is squeezed by gravity into a core that extends only a quarter of the way out from the centre to the surface and which occupies just 1.5 per cent of the volume of the Sun. The whole Sun is roughly 108 times the diameter of the Earth, which means that its volume is roughly a million times that of the Earth, and its mass is about 330,000 times that of the Earth. So a mass equivalent to 165,000 Earths is squeezed into a volume at the heart of the Sun only 25 times bigger across than the Earth.

In that core, electrons are completely stripped away from their atoms, leaving bare nuclei, and the resulting mixture of nuclei and electrons is squeezed to a density 12 times that of lead. The temperature is about 15 million K at the centre of the Sun, and drops to about 13 million K at the outer edge of the core. It is under these extreme conditions that a series of nuclear interactions takes place that converts hydrogen into helium. The multi-step process converts four hydrogen nuclei (also known as protons) into a single helium nucleus (also known as an alpha particle), which consists of two protons plus two neutrons. The crucial point is that each time this happens 0.7 per cent of the mass of the original four protons is released as energy, in line with Einstein's famous equation. That is what keeps the Sun shining, and why great efforts are being made to develop fusion reactors to mimic the process going on at the heart of the Sun to produce clean energy here on Earth.

Even under the conditions at the heart of the Sun, this process is quite rare. At the present rate, it would take about a hundred billion years to burn up all of the hydrogen in the core of the Sun in this way, although that will never happen, because of the buildup of helium 'ash' in the core. But there are so many protons inside the Sun that even though only a small proportion are converted into helium nuclei each second, that still corresponds to 5 million tonnes of mass being converted into energy and radiated away. In every second, 700 million tonnes of hydrogen is converted into 695 million tonnes of helium in the heart of the Sun. This process has been going on for about 4.5 billion years, since the Solar System formed, but has only used up about 4 per cent of the original stock of protons so far.

A star like the Sun, steadily burning hydrogen into helium, is said to be a member of the 'Main Sequence' of stars. How long a star can stay on the Main Sequence depends on its mass. Smaller stars are fainter and burn their fuel more slowly; larger stars are brighter and burn their fuel more rapidly. This is clearly relevant to the reason why we are here, since it has taken about 4 billion years for intelligence to evolve on Earth, and that would not have happened if the Sun had burned out a couple of billion years ago. But the Sun is roughly in the middle of the Main Sequence, and also roughly in the middle of its life as a Main Sequence star. Computer simulations and comparisons with other stars tell us that this steady process of converting hydrogen into helium inside the Sun can go on for about another 4 or 5 billion years.

Then the buildup of helium leads to a rearrangement of the core of the star, which will have profound effects for the Sun and for life on Earth that I describe in the next chapter.

During the next phase of its life, a star like the Sun 'burns' helium into carbon and oxygen. For the Sun itself, this is the end of the story, and when all of the helium fuel is used up it will settle down into a cooling ball of material, a stellar cinder about the size of the Earth, called a white dwarf. But more massive stars can take the process of nuclear fusion farther, because the density and temperature in the core of a star are both bigger for bigger stars. They can release energy by fusing nuclei together to make heavier and heavier elements all the way up to iron and nickel. Along the way, in the later stages of its life a star will swell up and blow away clouds of material into space, forming spectacularly beautiful subjects for astronomical photography but more importantly spreading the elements manufactured inside the stars out into space, where they can be recycled and form part of the material from which new stars and planetary systems form.

But the biggest stars do something even more spectacular. They explode, and in so doing they make all the elements heavier than iron.

Energy can be released by combining nuclei into heavier nuclei all the way from hydrogen up to iron, although there is a kind of law of diminishing returns which means that more energy is released (per particle involved) for the first step in the process, hydrogen to helium, and less and less for later step. The process stops at iron, because to make elements heavier than iron by fusing nuclei together energy has to be put in. To get energy out, instead of fusing nuclei together you have to split them apart – fission. This is the basis of nuclear fission reactors, where unstable nuclei of very heavy elements such as uranium and plutonium are encouraged to split into lighter elements, with energy being released. Those heavy elements only exist at all, and the energy is only available to be released, because energy was put in to their manufacture in the dying explosion of a star (or stars) that started out with at least eight to ten times as much mass as our Sun. The explosions are called supernovas, and the energy comes from gravity.

There are actually two kinds of supernova, but the first kind, known as Type I, does not manufacture any elements heavier than iron. Even so, they are interesting because they give an insight into what happens when a massive star dies. The fate of a star like the Sun is to end up as a white dwarf, a dense, inert lump of matter in which there is still a distinction between different kinds of atomic material, such as helium, oxygen and carbon, even though the atomic remnants are packed very closely together. But a straightforward calculation made originally by the Indian astrophysicist Subrahmanyan Chandrasekhar in the 1930s shows that if such a stellar remnant has more than about 1.4 times as much mass as our Sun, gravity will crush it into an even more compact state, in which all the atomic nuclei are squashed together into a ball of neutrons a few kilometres across.

To put this in perspective, most of the mass of an atom, more than 99 per cent, is contained in a tiny central core, known as the nucleus, composed of protons and neutrons – the essential difference between them is that protons have positive electric charge and neutrons have no charge. This nucleus is surrounded by a cloud of negatively charged electrons (the number of electrons in the cloud is the same as the number of protons in the nucleus). In very approximate terms, the size of the nucleus compared with the size of the entire atom is like a grain of sand compared with Carnegie Hall. This is why the changes in the structure of a star associated with supernovas are so dramatic.

The critical mass for a white dwarf is known as the Chandrasekhar limit. A Type I supernova occurs when a white dwarf that has a mass very near, but just under, the Chandrasekhar limit gains more matter from outside. This is quite likely to occur, since most stars are members of binary systems (this in itself is an important clue to why we are here), and a white dwarf in orbit around another star will tug matter onto itself from its companion because of its strong gravitational pull.

When the mass of the white dwarf reaches the Chandrasekhar limit, it collapses suddenly, from the size of the Earth down to the size of a mountain. As it does so, a wave of nuclear reactions runs through the material of the star, converting carbon and oxygen into iron. This process releases a huge amount of energy, with about half of the mass of the original white dwarf star being converted into iron, with traces of other elements such as sulphur and silicon, and scattered into space by the blast. The iron used to make the steel in your kitchen knives was made in this way. The remaining material becomes a ball of neutrons, with no remaining protons because electrons and protons are squeezed together to make more neutrons – a whole mountain-sized star with the same density as the nucleus of an atom. Most of the energy released in a Type I supernova explosion comes from the conversion of carbon and oxygen into iron. But most of the energy released in a Type II supernova explosion comes from gravity.

Stars which start their lives with more than about 8 or 10 solar masses of material cannot lose enough matter during their lifetimes to end up with less than the Chandrasekhar limit. They burn their nuclear material much more rapidly than the Sun does, because they need to hold a much greater weight up against the pull of gravity, and they take the nuclear-burning process much farther, all the way up to iron. For a star with 15 to 20 times as much mass as our Sun, the final stage of nuclear burning will be going on in layers around an inner core with as much mass as the Sun, roughly the size of the Earth, and mostly composed of iron. The core is like a white dwarf, but with more than 10 solar masses of material above it, held up only by the last phases of nuclear burning. Then the star runs out of fuel and nuclear burning stops. The entire weight of everything above the core presses down on the core, which collapses in about a tenth of a second all the way down to a neutron star (perhaps into a black hole), leaving a gaping hole beneath the 10 or more solar masses that formed the outer layers of the star.

It is the gravitational energy release by the total collapse of the inner iron core that provides the energy of a supernova. As the outer layers start to fall in, they are hit by a blast of energy and particles so powerful that it manufactures all the elements heavier than iron, and blows the material away into space where it mixes with the interstellar material from which new generations of stars and planets will be manufactured. The total energy released by a single Type II supernova is about a hundred times greater than the amount of energy that our Sun will release over its entire lifetime; for a few weeks a Type II supernova will shine as brightly as all of the other stars in its parent galaxy put together. The influence of supernovas is one of the most important reasons why we are here. And that influence is very different in different parts of our Galaxy.

MIXING METALS IN THE MILKY WAY

Clearly, the proportion of heavy elements in a star – its 'metallicity' – increases as the stellar generations follow one another. Since planets like the Earth are made of those heavy elements, the metallicity of a star should be a good guide to the chances of finding Earth-like planets in orbit around the star. The metallicity of a star can be defined in different ways, but the most common is to measure it in terms of the amount of iron the star contains, relative to the amount of hydrogen it contains.

Happily, measuring the proportion of different elements in a star is one of the most straightforward tasks in astronomy. When they are hot, the atoms of any particular element radiate light in very narrow bands of colour, called spectral lines. These correspond to precise wavelengths of light. Sodium, for example, radiates very brightly at two wavelengths of light in the yellow-orange part of the spectrum, which is why sodium street lights have their distinctive colour. The pattern of lines in the spectrum associated with each element is as distinctive as a bar code and can be used to identify the presence of the element unambiguously, by comparing the patterns we see in the light from stars with the patterns produced when different substances are heated in the laboratory. In a similar way, if white light containing all the colours of the spectrum shines through gas containing different kinds of atom, each kind of atom absorbs light at the same wavelengths that it radiates when hot, making equally distinctive dark lines in the spectrum. In either case, by comparing the strength of different lines in the spectrum – for example, iron compared with hydrogen – it is possible not only to tell which elements are present, but also to work out the proportions of the different elements present, without ever going near a star or a cloud of gas in space.

Metallicities are usually measured in terms of the metallicity of the

Sun, by comparing the ratio of iron to hydrogen in a star with the ratio of iron to hydrogen in the Sun. This is convenient, and it is hard to think of a suitable alternative, but a little misleading, since it is easy to slide from this into an unconscious assumption that the Sun is a completely ordinary or typical star. As I discuss in the next chapter, this is not necessarily the case. But sometimes simple assumptions are borne out by the observations. Common sense tells us that stars which contain more in the way of heavy elements ought to be more likely to have a family of planets, and the observations show that stars which are 'parents' to giant planets by and large do have higher metallicities than the average star in the nearby portion of the Milky Way. In particular, no giant planet has been found in orbit around any star with a metallicity less than 40 per cent of the Sun's metallicity. Until we actually observe large numbers of Earth-sized planets, we cannot be certain that high stellar metallicity is associated with other earths, but all the evidence points that way.

Too much metallicity may, however, be a bad thing. Giant planets, even bigger than Jupiter, have been found orbiting close in around stars with the highest metallicity, in orbits as near as, or nearer to their parent stars than, the Earth's orbit around the Sun. Nobody can yet say whether these massive planets formed in those Earth-like orbits, or whether they formed farther out from their parent stars and have migrated inwards as time has passed. But either way, the gravitational influence of such giant planets would disrupt the orbits of any Earth-like planets in the region, and send them either hurtling outwards or plunging inwards to a fiery doom. Either too little metallicity or too much metallicity seems to be bad news for the prospect of finding Earth-like planets in Earth-like orbits around other stars.

Because of how the Milky Way mixes up material, the time when a star is born is just as important as its location in determining its metallicity. I like to make the analogy with a pot of vegetable soup, kept simmering on a stove. Everyone who eats some of the soup adds something to the pot, so that it is never empty but gets richer and richer as time passes and more vegetables of different kinds are added to the mixture. The Earth, along with the rest of our Solar System, was born about 4.5 billion years ago, and among its many important characteristics our home planet has a large iron core, which, as we shall see later, has played an important part in maintaining conditions suitable for life on Earth, not least by generating a magnetic field which protects us from harmful radiation from space. Planetary systems that formed longer ago would have had proportionately less iron, so they could not have formed exactly Earth-like planets. Equally, planetary systems that are forming now, about 4.5 billion years after the Earth formed, are proportionately richer in iron. I'm not (yet) saying whether this is a good thing or a bad thing, as far as life is concerned, but it certainly makes those planets different from the Earth. Some planets are also likely to have proportionately less in the way of radioactive elements in their cores, and it is heat from radioactivity that keeps the core of the Earth molten today.

All of this is telling us that we have to consider our place in time in the Milky Way as well as our place in space, in order to understand the reason why we are here and why the Galaxy has not already been overrun by civilizations more advanced than our own. But our place in space is itself highly significant. The mixing of metals that is so important for the existence of planets like Earth and life forms like us only takes place in a – literally – very narrow part of the Milky Way.

The traditional classification of stars into 'populations' only has two categories – old, metal-poor Population II stars and young, metal-rich Population I stars. But more recent detailed studies show that stars in the Milky Way and other disc galaxies can more usefully be divided into four broad categories. Stars with distinctive ranges of ages and chemical compositions occur in four different parts of our Galaxy. The outer part of the visible Galaxy is made up of a sparsely populated halo of very old stars, at least twice as old as the Sun. The outer halo, at least, formed during the process which gave birth to our Galaxy from a collapsing cloud of gas about 10 billion years ago, although the inner part of the halo contains slightly younger stars and may have formed slightly later. This spherical halo is about 300,000 light years across, but contains very few stars. If life existed on any planets orbiting those stars, it would have had the opportunity for 5 billion years of evolution even before the Earth formed. It could have been as advanced as we are now when our ancestors were still single-celled bacteria – a seemingly dramatic demonstration of the Fermi paradox. But since very few of the halo stars have even 10 per cent of the Sun's metallicity, let alone the 40 per cent that seems to be necessary for planet-building, it is extremely unlikely that there are any Earth-like planets or life forms like us out there.

In contrast to the scarcity of stars in the halo, the most dense concentration of stars in our Galaxy is in the central bulge around which the disc has grown. It is hard to study stars in the bulge in detail, because dust in the disc of the Milky Way obscures our view in visible light, but infrared telescopes can probe beyond the veil of dust. Like the halo stars, many bulge stars are old and deficient in metals. But there is a wide range of metallicities among the bulge stars, and although very few, if any, are as young as the Sun or have anything like the metallicity of the Sun, at first sight this makes the region a slightly more promising location than the halo as a possible site for life. On the other hand, stars are so close together in the bulge and there is so much activity there (including outbursts associated with the black hole with 3 million times as much mass as our Sun right at the heart of the Milky Way) that cosmic radiation levels are likely to be too high for life to survive. Observations made by an unmanned space telescope called Integral have discovered energetic X-rays coming from an otherwise inoffensive cloud of hydrogen gas 350 light years away from the black hole at the heart of the Milky Way; the most likely explanation is that 350 years ago (as viewed from Earth) the black hole produced an outburst a million times more energetic than we see it today, and the radiation from that outburst reached the cloud 350 years after the event, making it fluoresce. Of course, all this 'really' happened some 27,000 years ago, since it has taken that long for the X-rays to travel out through the Milky Way to our telescopes. Either way, though, it is direct evidence that the bulge is not a good place for life.

That leaves the disc, which is itself made up of two components. There is a thick disc, about 100,000 light years across and 4,000 light years thick, which is made up of a population of old, metal-poor stars. Within this thick disc there is a much thinner layer, a thin disc which is also 100,000 light years across but no more than 1,000 light years thick –

the thickness is just 0.5 per cent of the diameter. If the disc were one metre across, its thickness would be just 5 millimetres. This very thin disc contains dust, gas, the Sun and young stars; it is the only region of the Galaxy where stars are still forming today, and metals are being mixed into a richer and richer soup.

It is straightforward to explain the differences between the thin disc and the thick disc. When a rotating cloud of gas collapses, it has no choice but to form a thin disc. Atoms and molecules of gas falling in from one direction collide with atoms and molecules of gas falling in from other directions, and in repeated collisions the random motions are cancelled out and everything settles into a disc. The same thing happens to clouds of gas as they settle into the disc. But stars are different. Stars passing through the thin gas of the disc are so big that they are scarcely affected by it, and the spaces between stars are so great that a star passing through the disc is extremely unlikely to collide with, or even pass near, another star. So a star can pass right through the disc before the gravity of all the matter in the disc pulls it back for another pass through the central plane of the Galaxy. The stars can pass to and fro repeatedly in this way, although gravity stops them getting too far away from the central plane. So the old, metal-poor stars of the thick disc are stars that were captured whole by our Galaxy when it swallowed up smaller galaxies. Any gas and dust swallowed up at the same time, however, had to settle into a very thin disc. So, like the stars of the halo, the stars of the thick disc are old, have low metallicities, and are highly unlikely to provide planetary homes for life. All that is left is the thin disc.

OUR PLACE IN THE MILKY WAY

New stars form in the spiral arms of the disc of the Galaxy. In ordinary photographic images, and especially in those made using blue light, the arms show up brightly, and it looks at first sight as if most of the stars in the galaxy are concentrated in the spiral arms. But in images taken in red light the arms are much less prominent, and it is clear that although there is a slightly greater concentration of stars in the arms, stars are spread more or less evenly around the disc of a galaxy, with the density of stars decreasing quite smoothly out from the centre of the disc to its edge. Spiral arms are bright because they contain a high proportion of bright stars, and that means they contain young stars.

The brightest stars are much more massive than the Sun, and blue-white in colour. But because they are so massive, they are short-lived, and burn out within about 10 million years. The blue-white stars tracing spiral arms must be young, because such stars never get old. They do not have time to move far from their birthplaces before they die, some of them in supernova explosions that enrich the interstellar soup.

Smaller stars, which are not so bright, are produced from the same collapsing clouds of gas and dust that form the bright stars, but they quietly go on their way around the Galaxy long after the large blue-white stars have faded away. Our Sun is just such a quiet member of the Milky Way community. Orbiting at a distance of about 27,000 light years from the centre of the Milky Way, at a speed of some 250 km per second, and taking about 225 million years to complete a circuit, the Sun and its family of planets have completed roughly twenty journeys around the Galaxy since the Solar System was born about 4.5 billion years ago. A massive blue-white star that formed in the same place and at the same time as the Sun would have completed only 5 per cent of a single circuit of the Milky Way before it exploded. At present we are near the inner edge (the inside of the curve) of a spiral feature known as the Orion Arm, or simply as the Local Arm.

But spiral arms are not permanent features of a disc galaxy like the Milky Way. They are concentrations of gas and dust where stars form, produced by disturbances within the Milky Way, or on occasions by a kick from outside, such as a gravitational tug from a nearby galaxy, or when the Milky Way swallows a smaller companion. Because of the way the Galaxy is rotating, any disturbance of its structure will be spread out in a spiral pattern. This is like adding a little cream or milk to a cup of black coffee that has already been stirred. The blob of white quickly gets spread out into a spiral pattern by the rotation. But almost as soon as the spiral pattern forms, continuing rotation winds up the spiral and smoothes it out of existence. Exactly the same thing happens in a disc galaxy after it is disturbed. The disturbance is spread out into a spiral pattern, which then gets smeared out and disappears. It's just that it takes hundreds or thousands of millions of years to do this in a galaxy, compared with the few seconds it takes to make and break a spiral pattern of cream in a coffee cup. The beautiful patterns of spiral galaxies only seem permanent to us because we are so short-lived; they are like snapshots of the spiral patterns in coffee.

In some cases, though, although the spiral pattern may change as time passes, it is constantly being renewed. This happens when there are repeated 'kicks' that stimulate the process. If the bulge at the centre of a galaxy is not exactly spherical, it can produce a gravitational influence that prevents the disc ever settling down completely. This can produce a wave of higher-density gas – a spiral density wave – that sweeps around the galaxy triggering star formation as it goes.

The process of star formation, once it gets started, can itself trigger repeated, if less regular, disturbances that sweep through a galaxy. This seems to be happening in the Milky Way. In regions where there is a lot of star formation going on, the hot, young stars radiate large amounts of ultraviolet energy into their surroundings, together with 'winds' of material escaping from the surfaces of the young stars, and the debris from supernova explosions. All of this squeezes nearby clouds of gas and dust, making them collapse and triggering another burst of star formation. The result is a feedback that produces a chain of star-forming regions that gets turned into a spiral arm by the rotation of the Galaxy. This process of self-sustaining star formation is very effective at mixing metals in the thin disc of the Milky Way.

A spiral density wave will be a region of star formation, but, curious though it may seem, the wave itself does not rotate at the same speed that the stars move around the Galaxy, although it does go in the same direction. The wave moves more slowly than the stars, so as the stars (and the clouds of gas and dust from which stars form) orbit round the Milky Way they repeatedly catch up with spiral arms and pass through them, just as we are about to pass through the Orion Arm. A neat example of what happens in such a situation can be seen if you turn on the tap above a kitchen sink, but leave the plug out so that water drains away. Where the water from the tap hits the surface of the sink, it forms a thin layer that spreads out in all directions. But at a certain distance from the centre (depending on how fast the water is coming out of the tap), the depth of the water increases in a step called a hydrostatic jump. The step stays the same, even though water molecules are constantly moving through it. When gas clouds moving round the Galaxy reach a spiral density wave, they pile up in the same sort of way that water piles up in a hydrostatic jump, and stars form when the clouds are squeezed. But stars born in previous encounters of this kind, like the Sun, pass through the density wave without noticing it is there.

They may, though, be affected by all the activity associated with density waves and spiral arms. Spiral arms are, after all, the places where supernovas occur. If a supernova exploded nearby, intense radiation would sweep across the Solar System, ripping into the Earth and killing off many living things. It would also damage the atmosphere of our planet, possibly destroying the ozone layer that protects us from damaging solar ultraviolet radiation and certainly changing the climate in ways that could only be detrimental to life forms adapted to the preceding status quo, whichever way the change occurred. A supernova occurring within 30 light years of the Solar System would destroy most life on the surface of the Earth.

Another, but less extreme, malign influence might come from the passage of the Solar System through one of the dusty gas clouds that congregate in spiral arms. In an extreme case, the heat and light from the Sun might be dimmed by intervening material, triggering an Ice Age. There are doubtless other hazards associated with the passage through what amounts to a cosmic traffic jam.

The last time the Solar System passed through a spiral arm was roughly 250 million years, or one orbit of the Milky Way, ago, at the start of its present circuit. The arm it passed through then was not, of course, the Orion Arm, because the arms themselves have been moving and changing as time passes; but the Sun was then roughly in the same part of its orbit that it is in now. Coincidentally – or was it a coincidence? – at that time the Palaeozoic era of geological time was brought to an end by one of the greatest catastrophes ever to hit life on Earth. A series of disasters wiped out 95 per cent of all marine life on Earth (not just individuals, but whole species). The Palaeozoic began about 570 million years ago, and had lasted for nearly 350 million years. It was a time that saw the development of fish in the sea, the emergence of life on to land, and the evolution of reptiles. But it was the death of so many species at the end of the Palaeozoic that opened the way for the survivors to evolve into new life forms, notably the dinosaurs, and to flourish.

One of the reasons why life on Earth found it tough at the end of the Palaeozoic is that continental drift had produced an arrangement of the land masses of the Earth which contributed to the development of an Ice Age. But such an extreme extinction event demands an extreme explanation, and although at such a distance in time we can never be sure, the circumstantial evidence points to the encounter with a spiral arm, even if we cannot identify the smoking gun itself. It's just as well our Solar System doesn't encounter spiral arms more often – but why doesn't it?

The reason why the Solar System hasn't encountered a spiral arm for so long is partly because of the distance we are from the centre of the Galaxy, which places us in a gap between arms, and partly because the Sun's orbit around the Milky Way is, unusually, very nearly circular. The Sun has stayed in this gap between the arms for a long time because, although the whole Solar System orbits the Galaxy once every 250 million years or so, the spiral pattern takes twice as long to move around the Galaxy. By the time the Sun has completed one orbit, the pattern has moved on by half an orbit, so it takes correspondingly longer to catch up, giving a long time for evolution to do its work before being interrupted. Even in a circular orbit, a planetary system closer to the centre of the Milky Way would encounter spiral arms more often, because the arms wind up towards the galactic centre. Planetary systems farther out from the middle of the Galaxy than ourselves might encounter spiral arms even less often than we do; but there are good reasons to think that there may be few, if any, planetary systems out there.

THE GALACTIC HABITABLE ZONE

Since the diameter of the disc of the Milky Way is about 100,000 light years, the distance from the centre of the Galaxy to the outer edge of the thin disc is about 50,000 light years. The Sun is about 27,000 light years from the centre, a little more than halfway to the edge of the disc. Spectroscopy reveals that, by and large, stars that are closer to the centre of the Galaxy contain more metals, and there are many very old stars in the bulge. This is typical of disc galaxies in general, and supports the idea that they grew from the centre outwards. Although star formation continues at a modest rate today throughout the thin disc, there seems to have been a wave of star formation in our own Galaxy which started in the heart of the Galaxy and spread out from the centre, reaching the radius where the Solar System later formed between 8 and 10 billion years ago before spreading into the outer parts of the disc. But even after this wave had passed, it still took time, and further generations of stars, to build up the metallicity at our distance from the centre of the Milky Way to the point where a star like the Sun could form. Farther out, even 5 billion years ago the metallicity had not reached this level. Because there is a close link between the metallicity of a star and the likelihood that it has planets, this has led to the idea of the 'Galactic Habitable Zone' (GHZ), the region around the thin disc where planetary systems like the Solar System and planets like the Earth are likely to be found. The crucial point, as far as the prospect of other intelligent beings existing in our Galaxy is concerned, is that this Galactic Habitable Zone slowly gets larger as time passes. According to some calculations, it also moves outwards through the disc as time passes; but the most up-to-date version of the idea, from Charles Lineweaver and his colleagues, suggests that it has always been centred on a ring about 26,000 light years from the galactic centre, began to emerge about 8 billion years ago, and at present extends from about 23,000 light years to about 29,000 light years. The Sun is close to the centre of the GHZ, but not exactly at the centre. This is the region where the abundance of metals 5 billion years ago, when the Solar System formed, was sufficient to allow for the formation of planets like the Earth.

At any distance from the centre of the Milky Way, the metallicity of the gas from which new stars and planetary systems can form is still increasing as time passes. But farther from the centre there is relatively less gas and less star formation, so even today the metallicity in the outer regions is not increasing as rapidly as in the inner regions. Because metallicity is measured in terms of the metallicity of the Sun, you might expect that the metallicity of the Sun itself is 1, by definition. So it is. But because of the rate at which metallicity falls off with distance from the galactic centre, astronomers prefer to use logarithmic units, which are more convenient for comparing a wide range of values. The logarithm of 1 is 0, so on this scale the metallicity of the Sun is defined as 0, in logarithmic units astronomers call 'dex'. The nomenclature doesn't matter, but what does matter is that on this scale a negative number simply means that a star has a smaller metallicity than the Sun, not that it has less than zero metals!

At the distance of the Sun from the centre of the Milky Way, the metallicity today goes down by a little more than 5 per cent for every additional thousand light years from the centre. In logarithmic terms, the decline is 0.02 dex per thousand light years. Other disc galaxies show very similar metallicity gradients. I shall have more to say about why stars that are over-rich in metals might not possess Earth-like planets later, but the position of the GHZ also depends on the kind and frequency of hazards that a potential home for life encounters on its journey around the Milky Way. I have already mentioned the radiation risk from supernovas. The peak of supernova activity among the stars of the Milky Way seems to be at about two thirds of the Sun's distance from the centre of the Galaxy. But the radiation hazard comes not only from exploding stars but also from the centre of the Milky Way itself, where there is a large black hole that is quiet today but shows signs of having been active in the recent past. This black hole contains about 3 million times as much mass as our Sun, within a volume no bigger across than forty times the distance from the Earth to the Moon. Studies of active black holes in other galaxies, combined with our understanding of the physics of black holes, tell us that when the black hole swallows matter, which happens when a star or gas cloud gets too close to it, the material funnelling into this small volume at high speed gets very hot and radiates intense amounts of electromagnetic energy (such as X-rays), while jets of electrically charged particles are shot out into the bulge. This would be bad enough for stars in circular orbits around the centre; but stars there tend to be in very elliptical orbits that take them, and any associated planets, diving in close to the site of all this activity.

Explosions far more powerful, but far less common, than supernovas also occur in galaxies like our own from time to time, and are also more likely to occur near the centre of a galaxy, where the density of stars is greatest. These explosions are revealed by the intense but short-lived gamma ray bursts that they emit, and are known as gamma ray bursters. The exact mechanism which powers these bursts is still a mystery, but a gamma ray burster is the most powerful explosion that occurs in the Universe today, emitting more energy in a few seconds than the Sun will emit in its lifetime. They can be seen far away across the Universe, and the statistics suggest that such a burst occurs in a galaxy like the Milky Way roughly once every hundred million years. Even if one occurred in a distant region of the Milky Way, the blast of sterilizing radiation could be devastating for life on Earth (or any other inhabited planet in the Galaxy) and destructive to the ozone layer. Conceivably, a strong gamma ray burst could sterilize an entire galaxy. More optimistically, though, it has been pointed out that since such bursts last for less than a minute, only the side of the planet facing the burster will be affected. The other side could be sufficiently shielded for life to survive, even if it suffers a setback. It is possible that one reason why it has taken so long for intelligent life to emerge on Earth (and perhaps elsewhere in the Galaxy) is that gamma ray bursters were more common when the Galaxy was young, and it is only in the past few billion years that life has had a chance to evolve to the point of producing at least one technological civilization.

Putting all of the evidence together, Lineweaver and his colleagues conclude that the GHZ contains no more than 10 per cent of all the stars ever formed in the Milky Way. But there are other hazards within the GPZ.

CATASTROPHIC COMETS

Compared with the extreme outpourings of a black hole, a comet might seem a minor threat. So it is, for a planet – but not for life. The problem is that cometary impacts, like the one which happened on Earth 65 million years ago at the time of the death of the dinosaurs, can wipe out complex life forms on a planet at a stroke. If such events happen at intervals of tens or hundreds of millions of years, life can cope (curiously, as we shall see, life may even benefit, in the long term). But frequent cometary impacts will prevent life having time to evolve intelligence, if species are wiped out too often.

Comets are lumps of ice and rock left over from the formation of the Solar System. The familiar photogenic comets that we all know, if only from pictures, with their long tails stretching out behind them, are actually rare visitors to the inner part of the Solar System from a vast cloud of comets that surrounds the Solar System, like an eggshell surrounding the yolk of an egg, almost halfway to the nearest stars. This is known as the Oort Cloud, after Jan Oort, an astronomer who studied comets and calculated the properties of the shell. The bright tails that comets grow as they approach the Sun are caused by the heat from the Sun vaporizing the icy material to release gas and dust; and the ices themselves do not only consist of water but include things like frozen methane and ammonia.

But the threat to life on Earth has nothing to do with the composition of comets. If you are hit on the head by a one tonne lump of ice, it does just as much damage as being hit on the head by a one tonne lump of rock. A lump of ice and rock 10 km across striking the Earth at a speed of 50 km per second would release energy equivalent to the explosion of a hundred million megatonnes of TNT – more than 5,000 million times the energy released by the atomic bomb dropped on Hiroshima at the end of World War Two. This would be more than enough to explain the global environmental disruption that occurred at the time of the death of the dinosaurs. From counting the scars of old craters on the surface of the Earth and other planets, and our knowledge of the orbits of comets and asteroids, astronomers estimate that an impact like this occurs on Earth roughly every hundred million years. As the fate of the dinosaurs shows, this poses severe problems for the evolution of life on Earth, although – as our existence testifies – such a catastrophe may open up new opportunities for the survivors of the disaster to spread and diversify. If the impacts occurred a lot more frequently, though, it seems likely that there would never be time for intelligence to evolve in the interval between catastrophes. And impacts are likely to be much more common on planets orbiting stars closer to the galactic centre than we are.

The reason is that comets are shaken loose from the Oort Cloud by the gravitational influence of any stars or clouds of gas that the Solar System encounters on its journey round the Milky Way, and then fall into the inner Solar System where they pose a threat to life on Earth. Our understanding of the way the Solar System formed tells us that comet clouds like the Oort Cloud will be a feature of all planetary systems like the Solar System, so they will all be subject to the same kind of hazard. So how big can the risk be? Pretty big. It is estimated that the Oort Cloud contains several thousand billion comets, although the total mass of all the comets in the cloud put together would be only a few tens of times the mass of the Earth. Close encounters of the kind that will shake comets loose from the cloud will be much more common closer to the centre of the Galaxy, where the stars are closer together; they will also be more common when a planetary system is crossing one of the spiral arms, where gas clouds pile up in the equivalent of a hydrostatic jump.

The exact boundaries of the GHZ are not clear, but what is clear is that the inner regions of the Galaxy have plenty of metals but are hazardous for life, while the outer regions of the thin disc are safer, but metal-poor and unlikely to contain Earth-like planets. In between, there is a Goldilocks region, the GHZ, that is just right for life. The Solar System sits near the centre of that zone. We live in a particularly favourable place in the Milky Way for life. We also live at a particularly favourable time in the Milky Way for life. Until about 5 billion years ago, even apart from the shortage of metals, the activity of starbirth and the supernovas associated with stardeath would have made life hazardous. It has taken almost all of that 5 billion years for intelligence to evolve on Earth, and if (what a big if!) that is typical, apart from any other consideration we may be one of the first, if not the first, intelligent civilization in our Galaxy. There is indeed something special about our place the Milky Way, in both time and space.

Incidentally, the existence of the Galactic Habitable Zone reinforces the power of the Fermi paradox. If 'they' exist anywhere in the Milky Way, they will live in the GHZ. And if they want to explore the Milky Way looking for other life-bearing planets, they would only have to explore the GHZ, not the whole of the Galaxy. Any spacefaring civilization would surely know enough about the stars to realize this. Which makes the task of sending robot probes to visit every possible home for life even easier, and quicker!

Although the discussion in this book is focused on the Milky Way Galaxy, it is worth mentioning that the prospects for life in other galaxies are even gloomier, as far as life forms like us are concerned. In the relatively nearby regions of the Universe that we can study in detail with our telescopes, four fifths of all stars are in galaxies that are intrinsically fainter than the Milky Way. This faintness is an indication that the processes of starbirth and stardeath that enrich the chemical soup in those galaxies are weaker, and they have lower overall metallicities than the Milky Way. The Milky Way itself may be in a minority, about 20 per cent of the total, of inhabitable galaxies; and the GHZ contains a minority, at the very most 10 per cent, of the stars of the Milky Way. But how many of those stars are likely to have planets on which life has evolved? Is the Sun itself special compared with its neighbours in the Galactic Habitable Zone?

3
What's So Special about the Sun?

Just as there is a habitable zone around a galaxy like the Milky Way, so there is a habitable zone around a planet like the Sun. The essential requirement of this Stellar (or Solar) Habitable Zone, abbreviated as SHZ, is that it covers the region around a star where liquid water can exist – it is neither so cold that water freezes, nor so hot that water boils. At first sight, this may seem unduly restrictive. Science fiction writers have imagined life existing on planets with oceans of liquid methane, and in other exotic environments. But water has unique properties that make it an ideal medium in which life can evolve.

The most important reason why water is (literally) vital for life is that life needs a solvent – a medium in which chemicals dissolve and chemical reactions can take place. Water is far and away the best liquid for this job; ammonia comes closest, but lacks the other special properties of water.

The second most important property of water for life is that each molecule has a magnetic polarity. In other words, one end of a water molecule behaves like a very weak magnetic north pole, and the other end like a very weak magnetic south pole. This is an almost unique property in the world of molecules. This polarity encourages not only water molecules but molecules dissolved in water to line up in certain ways, and this is one factor in determining the shape of the amino acid molecules that are crucial for life.

A third property of water is, when you think about it, quite bizarre, although it can be summed up in two words. Ice floats. In other words, solid water is less dense than liquid water. This is because when water freezes the molecules line up to form a very open crystalline structure, where the molecules are farther apart than they are when they are rubbing shoulders in liquid water. In other substances, the molecules are closer together in the solid, so it is more dense than the liquid. The physical and chemical reasons for this strange property of water are well understood, but are not important here. What is important is that this means that during an Ice Age ice forms a skin on top of the ocean at high latitudes, keeping warmer water beneath this insulating layer. If the ice sank to the bottom of the ocean, the top layer of water would be uncovered and freeze in its turn, with the process repeating until the oceans were a solid ball of ice.

All in all, it is reasonable to assume that 'life as we know it' does require the presence of liquid water. So how much does that restrict the SHZ?

THE NARROW ZONE OF LIFE

Some people have argued that even restricting the SHZ to the region where temperatures are between 0 °C and 100 °C is too generous, and that we should only look at the zone from 0 °C to 50 °C. Their argument is based on the fact that complex life forms like ourselves cannot survive at temperatures above about 50 °C. This is certainly true of animal life on the surface of the Earth; but in recent decades (beginning in the late 1970s) studies of the deep ocean floor have revealed the existence of hydrothermal vents, like underwater geysers, that teem with a variety of life forms that feed off the energy and chemicals released by the vents. These vents are thousands of metres below the surface of the sea, where no sunlight ever reaches, but support complete ecosystems including eyeless shrimps and so-called 'Pompeii worms' (named after the volcano) which live in tubes where the water temperature exceeds 80 °C. Even complex life, it seems, can cope with temperatures well above 50 °C, provided it has a supply of liquid water. So it is best to consider the life zone around a star as covering all the region from 0 °C to 100 °C – which is still restrictive enough to give further evidence that we occupy a special place in the Universe.

When we look at the Solar System today, we see that the Earth is almost in the middle of the SHZ. Venus, the next planet in towards the Sun, is too hot for liquid water to exist; Mars, the next planet out from the Sun, is too cold. But the life zone has not always been in the same place. Our understanding of the processes which keep stars hot, and comparisons with others stars, tell us that the Sun was cooler when it first formed and has been steadily warming up as it has aged. Roughly speaking, 4 billion years ago the Sun was 25–30 per cent cooler than it is today. So over the past 4 billion years, the life zone around the Sun has steadily moved outwards. The region that was on the outer edge of the zone (the cool edge) is now on the inner edge (the hot edge), and regions that used to be in the hottest part of the zone, including the orbit of Venus, are now too hot for life. This has led to the concept of the Continuously Habitable Zone, or CHZ, which is the region around the Sun where the temperature has always been between 0 °C and 100 °C. This is sometimes known as the 'Goldilocks Zone', because the temperature there, like the temperature of baby bear's porridge in the story, is 'just right'. The CHZ is tiny. It extends from a distance just 1 per cent farther out from the Sun than we are to a distance just 5 per cent closer to the Sun than we are.

It looks as if the Earth's orbit is in a quite extraordinarily lucky part of the Solar System, as far as the prospects for the evolution of intelligence are concerned. But the situation isn't quite as clearcut as it seems at first sight. The presence of life on Earth plays a part in regulating the temperature of our planet, through the greenhouse effect. Gases such as carbon dioxide act to warm the surface of the Earth by trapping heat that would otherwise escape into space. Today, this natural greenhouse effect keeps the Earth about 33 °C warmer than the surface of the airless Moon, even though the Earth and the Moon are at very nearly the same distance from the Sun. When the Earth first formed, the atmosphere was richer in such greenhouse gases, preventing it from freezing even though the Sun was cooler. As the Sun warmed and life developed on Earth, carbon dioxide was drawn down out of the air by living things and deposited as carbonate rocks, reducing the strength of the greenhouse effect. Life alters the amount of carbon dioxide in the air, through feedback processes which keep the planet warm when the Sun is cool and prevent it overheating as the Sun warms up. This is the basis of Gaia theory, developed by James Lovelock, which I discuss later.

One other example highlights the importance of feedbacks to the location of the CHZ. Weathering of rock involves chemical reactions which also remove carbon dioxide from the air. When the planet is warmer, more water evaporates from the seas to fall as rain, and weather systems, driven by convection, are more intense, so there is more weathering. This removes carbon dioxide from the air, weakening the greenhouse effect and cooling the globe. But when the planet is cool there is less weathering, so carbon dioxide (released from volcanoes) builds up in the air, strengthening the greenhouse effect and thereby warming the planet.

All of this extends the boundaries of the CHZ around the Sun, but not, in all honesty, by much. It hardly changes the inner boundary of the CHZ at all, leaving it just 5 per cent closer to the Sun than we are; but it pushes the outer boundary out to 15 per cent farther from the Sun than we are. In units where the radius of the Earth's orbit is defined as 1 astronomical unit (AU), the CHZ extends from 0.95 AU to 1.15 AU; so it spans a distance equivalent to 20 per cent of the size of the Earth's orbit. It's just as well that the Earth is in a nearly circular orbit around the Sun; in fact, our orbit deviates from a perfect circle by only 1.7 per cent. If our planet were in an elliptical orbit as extreme as that of Pluto, at some times of the year it would be closer to the Sun than the inner edge of the CHZ, and at other times of the year it would be farther out than the outer edge of the CHZ. Elliptical orbits are bad news for life – a theme I shall return to. But as these limits indicate, even in a nice circular orbit Earth's days as a home for life are numbered. By the time the CHZ has crept outwards by another 5 per cent, which will happen in a couple of billion years from now, the Earth will become like Venus and be too hot for life. So the habitable lifetime of the Earth is about 6 billion years. How does this compare with the CHZs for other stars?

THE SUN IS NOT AN AVERAGE STAR

Our Sun is often described as being an average star. This is only true in a very narrow sense. Stars that, like the Sun, are maintaining their output of energy by converting hydrogen into helium deep in their interiors are said to lie on the 'main sequence' of a kind of graph, called the Hertzprung-Russell Diagram, which astronomers use to relate the temperature of a star to its mass. Because the Sun is on the Main Sequence of this diagram, it is regarded as an ordinary star. But ordinary does not mean average. Some 95 per cent of all stars are less massive than the Sun, and because the brightness of a star is related to its mass, this means that they are dimmer than the Sun. In that respect, the Sun is far from being average, and stars that are bigger and brighter than the Sun are even more rare than stars with the same mass as the Sun, even though massive stars are quite normal.

The Sun may not be 'average' in another way. There is some evidence that the brightness of the Sun varies by less than the variation in brightness of other stars with similar masses and chemical compositions. This is very hard to quantify, and we cannot be sure whether this has always been the case or is just a phase the Sun is going through today (or for the past few million years). But it does at least hint that the Sun may be an unusually stable star, with obvious benefits for the evolution of life on Earth.

On a larger scale, being either brighter or dimmer than the Sun has dramatic implications for the CHZ around a star. Most of the stars in the Galaxy – 95 per cent – are smaller and fainter than the Sun. Three quarters of all the stars in our neighbourhood are so-called red dwarfs, a category also known as M-type stars, which have only about a tenth as much mass as our Sun. Red dwarfs live for much longer than stars like the Sun (which is a yellow-orange G-type star; the initials are a historical accident and have no significance except as labels). This would be a good thing in terms of allowing time for intelligence to evolve. Unfortunately, though, the conditions on any planet orbiting a red dwarf are likely to be unsuitable for the emergence of a technological civilization.

The first problem is that the life zone around a red dwarf is very narrow, and very close to the parent star. In order to have liquid water on its surface, a planet would have to orbit within 5 million km of the star, at a distance only one thirtieth of the distance of the Earth from the Sun. Even at its closest, Mercury, the innermost planet in our Solar System, never gets within 46 million km of the Sun. It isn't clear that planets could even form, or occupy stable orbits, within 5 million km of a star, but even if they could there would be complications. Just as tidal forces have locked the Moon into a rotation which keeps one face always turned towards the Earth, so planets in the life zone around a red dwarf would be locked into a rotation with one side always facing the star. So one side would be in eternal darkness, and the other in eternal light. Except, possibly, for a narrow twilight zone, the conditions would be either uncomfortably hot or uncomfortably cold. The most likely consequence of this is that convection would carry gases from the hot side of the planet to the cold side, where they would cool and freeze. Any atmosphere the planet originally possessed before the tidal locking was completed would freeze out on the dark side.

Another problem – as if that weren't enough – is that red dwarf stars are much more active than the Sun. They produce frequent flares of activity which release large amounts of ultraviolet radiation, X-rays and particles. This would be particularly damaging because of the proximity of the planet to the star. Apart from the direct consequences for life, these outbursts would strip away any atmosphere that started to form around the planet. Overall, it seems we can rule out red dwarf systems as likely homes for other civilizations. We have already found that the Galactic Habitable Zone only includes 10 per cent of the stars in the Milky Way, and now we are ruling out 75 per cent of that 10 per cent. That leaves us with only 2.5 per cent of all the stars to consider, and we have barely started identifying all the reasons why we are here on Earth.

Bigger, brighter stars than the Sun form only a small part of that 2.5 per cent, and in terms of habitable zones alone are no better than red dwarf stars as possible places to find planets harbouring technological civilizations. A brighter star has a larger habitable zone, but it doesn't live as long as the Sun, and the habitable zone moves out more rapidly than the Sun's habitable zone as the star ages. A star with 30 times as much mass as our Sun would have to burn its nuclear fuel so fast that the rate at which it pours out energy is 10,000 times that of the Sun, and it will live for only a few tens of millions of years on the stable Main Sequence. Such stars also emit large amounts of ultraviolet radiation, damaging both to life and to the atmospheres of prospective Earth-like planets. The brightest stars on the Main Sequence, known as O and B stars, together make up less than one tenth of 1 per cent of all stars, though, so taking them out of the equation hardly makes much difference.

Slightly smaller, cooler A-type stars could provide any planets in their life zones with a stable environment for about a billion years, which is certainly long enough for life to get started, judging by the example of the rapid establishment of life on Earth, but may not be enough for a civilization like ours to develop. Even a star with just 1.5 times the mass of our Sun would leave the Main Sequence after only a couple of billion years. But there are some stars, the F-types, which are a little more massive than our Sun, have Main Sequence lifetimes of about 4 billion years, and which don't seem to produce excessive amounts of ultraviolet radiation.

Putting everything together, reasonably large, reasonably long-lasting life zones may exist around stars which are in the Galactic Habitable Zone and are like the Sun (G-type), or stars a little more massive (the F-types) or a little less massive (known as K-types). A generous assessment would make that no more than 2 per cent of the stars in the Galaxy. In that sense, we can already see that the Sun is special. But even within that 2 per cent, the Sun is not an average star, because most stars have companions – they live in binary or even triple star systems.

PERTURBING PARTNERS

It is actually very difficult to make stars. The large clouds of gas and dust in the thin disc of the Milky Way (known as giant molecular clouds, because they are big and contain molecules) rotate, which tends to stop them collapsing, and are threaded by magnetic fields which also help to hold them up against the inward tug of gravity. If a star with the same mass as the Sun formed from a cloud spread out to the density of a slowly rotating interstellar cloud, by the time this had shrunk to the size of the Sun it would be spinning so fast that its surface would be moving at 80 per cent of the speed of light. This is because a property known as angular momentum is conserved when a cloud shrinks – or, indeed, when it expands. In order to have the same angular momentum, provided it has the same mass a small object has to spin faster than a large object. This is exactly why a spinning ice skater can spin faster or slower by pulling their arms in or out. In order to shrink, a collapsing cloud of gas has to get rid of angular momentum. If two or more stars form from the same collapsing cloud, a lot of the angular momentum goes in to the orbital motion of the stars around each other, rather than into the spin of the stars themselves.

An average giant molecular cloud is about 65 light years across and contains about a third of a million solar masses of material. When a cloud passes through the density jump of a spiral arm, it gets squeezed, and if a supernova explodes nearby shock waves from the blast will go rolling through it. Under these conditions, turbulence stirring up the cloud can produce regions of greater density where gravity can take over and cause some of those local regions to collapse to form stars. Stellar 'nurseries' where this process is going on have been photographed in the infrared part of the spectrum, where radiation penetrates the dust in the clouds, from unmanned space observatories such as Herschel, confirming astronomers' understanding of what goes on based on their knowledge of the laws of physics.

Turbulence seems to produce 'pre-stellar cores' on which stars grow as gravity tugs more matter towards them. A typical core would be about a fifth of a light year across, and contain about 70 per cent as much mass as the Sun. Only the very centre of such a core collapses and heats up to the point where it begins to generate energy by nuclear fusion, becoming initially a tiny proto-star with less than a hundredth (perhaps as little as a thousandth) of the mass of our Sun; the nuclear reactions begin when it has grown to about a fifth of the mass of the Sun. The final size of the star that grows onto this core doesn't depend on the size of the core – all such cores start out with roughly the same mass. What matters is the amount of matter close enough to be captured by the gravity of the young star, before the radiation from the star and any companions forming nearby disperses the clump in the giant molecular cloud from which they have formed. For a star like the Sun, 99 per cent of its mass is gathered in this way by accretion. But this is a very inefficient process. Although roughly half of the mass in a clump gets turned into stars, only a few per cent of the material in the whole cloud is converted into stars as it makes the passage through a spiral arm.

Because of the angular momentum problem, it is hard to see how a star could form in isolation, and observations of our stellar neighbourhood show that at least 70 per cent of Sun-like stars have at least one companion, although systems with more than three stars bound together by gravity are extremely rare. Computer simulations of the way stars in multiple systems interact with one another and with nearby systems explain how this proportion has arisen, and why there are at least some stars which, like our Sun, do not have a stellar companion.

When three stars are orbiting around one another, they follow a complicated dance in which it is quite easy for one of the stars to gain a lot of energy and be ejected from the system, carrying angular momentum off with it, while the other two move closer together in a tighter embrace. Binary pairs are more stable, unless they pass close by another star (or a binary or a triple), in which case gravitational interactions can break up the pair and leave at least one isolated star, although its companion can sometimes be captured by the other system. Computer simulations suggest that if out of every 100 new star systems 40 are triple and 60 are binaries (making a total of 240 stars) then, allowing for how close these systems are in the star-forming regions of the Milky Way, by the time the star systems have moved apart into the Galaxy at large and things have settled down there will be 25 triples, 65 binaries and just 35 single stars. The same 240 stars are now shared out in such a way that just under 20 per cent are unaccompanied, roughly matching our observations of the stars in our neighbourhood.

Binary and triple star systems are bad news for life – certainly for the prospects of a technological civilization arising on any planet in such a system. Stable orbits can exist, either if the two stars in a binary are very close together (within about a fifth of the distance from the Earth to the Sun) and the planets orbit around both stars, or if the two stars are far apart (at least 50 times the distance from the Earth to the Sun) and the planets orbit one of the stars. But although the orbits may be stable, they will not be as beautifully circular as the Earth's orbit around the Sun, and the planets will be affected by the heat and light from two stars, making it difficult to establish a long-lasting habitable zone. Judging by the evidence of the geological record of the evolution of life on Earth, even a change in the amount of heat reaching a planet from its star or stars of 10 per cent could cause severe problems. A rough rule of thumb is that a 1 per cent change in the output of the Sun causes a 1 °C change in the average temperature at the surface of the Earth, and there is serious concern today that a global warming of 4–5 °C could cause the collapse of civilization.

Compounding these problems, in a binary system instead of having a single star getting steadily brighter with a well-defined habitable zone moving outwards as time passed, we would have two stars each getting brighter, at two different rates, to complicate the picture. The habitable zone (or zones) would move more rapidly, and in a less straightforward fashion. This might not be so bad for the kind of single-celled life forms that have existed on Earth almost since the Solar System formed, but it would not provide the long-term stability needed for the development of a technological civilization.

Apart from the difficulty of finding a long-lasting habitable zone in a multiple star system, it may be difficult for planets to form in such systems anyway, since the dusty discs of material from which planets form are likely to be disrupted by the complex tidal forces at work in such systems.

So only 30 per cent of the 2 per cent of the stars of the Milky Way that are both in the GHZ and roughly Sun-like are left for our consideration. That is, just 0.6 per cent of all the stars in the Milky Way. If the existence of planets actually requires a star that does form completely in isolation by some rare and not yet understood process, that tiny proportion reduces further by a large but unknown factor. And we still haven't exhausted the list of properties that make the Sun special.

BLASTS FROM THE PAST

Astronomers know a great deal about the rough-and-tumble conditions in which the Sun formed, because events occurring at that time have left traces in the form of radioactive elements found in the Solar System. Like all the elements except hydrogen and primordial helium, these radioactive elements were made inside stars and blown out into the material from which new stars and planets form when their parent stars die. But radioactive elements do not last for ever, and they are only formed in large stellar explosions. So we can say with certainty when they were formed, and that they were formed in an exploding star or stars close to the site where the Sun formed.

Each radioactive element decays into a stable substance, sometimes in a multi-step process, on a timescale known as the half-life. This half-life is different for each element (strictly speaking, for each isotope of the element; atoms of different isotopes have the same chemical properties but different weights). The stable elements produced by this process are known as the daughter products of this element – radioactive uranium, for example, decays ultimately to produce lead. Comparing the proportion of a radioactive element found in a sample of material today with the proportion of its daughter products in the same sample reveals how much of the radioactive material has decayed, and this tells us just how long it has been since the radioactive material was formed.

Two of the radioactive isotopes produced in the death-throes of a star are iron-60 and aluminium-26. Iron-60 has a short half-life, soon decaying into its daughter isotope nickel-60, but aluminium-26 decays much more slowly. If they had both been manufactured at the same time in a star that exploded and laced the Solar System with radioactive debris, then there would be a certain proportion of nickel-60 compared with aluminium-26 in the oldest pieces of the Solar System.

But when researchers from the University of Copenhagen studied old rocks from Earth and samples of meteorites, rocks from space thought to be leftover pieces from the formation of the planets, they found less nickel-60 in the oldest meteorites and more in slightly younger meteorites. A plausible explanation of this puzzle is that a very large star, perhaps 30 times as massive as our Sun, blew away its outer layers, including a lot of aluminium-26, about a million years before the final explosion of the remaining core of the star. This initial outburst would have been enough to trigger the collapse of the knot in the giant molecular cloud from which the Solar System formed. Then after a million years or so the star exploded, showering its surroundings, including the forming Solar System, with debris, including iron-60 dredged up from deep in its interior.

In order to do the trick, the supernova explosion must have occurred within a tenth of a light year of the forming Solar System, when the Sun was less than 2 million years old. No other supernova has ever exploded in such close proximity to the Sun – if it had, life on Earth would have been exterminated. It cannot be a coincidence that this seemingly unlikely event occurred where the Solar System was forming, and the natural explanation is that both the Sun and the supernova were members of a cluster of stars that formed together in the same gas cloud, and have since gone their separate ways.

There are good examples of clusters like this. One of the best-known is called R136, and is located in the Large Magellanic Cloud, a small companion galaxy to the Milky Way. The cluster contains about 10,000 stars, none of them more than a few million years old. Clearly, the stars formed together out of a large cloud of gas and dust. Because massive stars are rare (but rare in a quantifiable way, compared with the number of stars like the Sun), we can set limits on the size of the cluster in which the Sun was born. Roughly speaking, there is one star with 30 times the mass of the Sun for every 1,500 smaller stars, so there must have been at least that many stars in the cluster where the Solar System formed. On the other hand, large clusters take longer to settle down into a compact state where stars are as close as the Sun was to the supernova that spread radioactive debris across the Solar System, and a star with 30 times as much mass as the Sun lives for only a few million years before exploding. Dutch astronomer Simon Zwart calculates that on this basis the cluster in which the Solar System formed cannot have contained more than about 3,500 stars. This is small as clusters go, and the gravity of the stars in the cluster tugging on each other would not be strong enough to prevent the cluster being disrupted and its constituent stars spread apart by tidal forces and interactions with other objects that it encountered as it moved around the Galaxy. It would have been completely disrupted within 250 million years, by the time it had completed one circuit of the Milky Way.

But for a brief time, while the Solar System was forming, the stars in the cluster would have been so close together that some would pass closer to one another than Pluto is to the Sun. This would not be good news for any planets trying to form around the stars. More of this shortly; but first there is another puzzle, about the composition of the Sun itself, that seems to make our place in the Universe special.

THE MYSTERY OF SOLAR METALLICITY

The metallicity of the Sun can be inferred in two ways. The first is to measure the proportions of different elements present in the surface of the Sun, using spectroscopy. As we have seen, each element produces a characteristic pattern of lines in the spectrum of light from the Sun, and the strength of those lines indicates the amount of each element present. The other technique probes the Sun's interior using helioseismology. As its name suggests, this is like the way geophysicists probe the interior of the Earth using seismology. Sound waves inside the Sun make the surface vibrate, and these vibrations can be measured by sensitive instruments. The nature of the vibrations tells astronomers about the density of the solar interior, and the speed of sound at different depths inside the Sun. These properties depend on the exact composition of the Sun.

Until recently, both techniques seemed to be telling us that the metallicity of the Sun is unusually high compared with the average metallicity of stars at our distance from the galactic centre. This is particularly puzzling because Zwart's orbital calculations suggest that if anything the cluster of stars in which our Sun was born formed a little farther out from the centre than we are now. The obvious explanation is that the supernova which enriched the Solar System with iron-60 also enriched the Sun with other heavy elements. That would make it much easier for the Sun to form a family of rocky planets than it would be for stars in our neighbourhood that have not been enriched in this way.

Recently, though, new observations of the solar spectrum have been interpreted as indicating a lower metallicity than the older observations suggested. These claims have not been unambiguously confirmed, but even if they are right nothing has changed the helioseismology evidence, so the implication would be that the Sun is more metal-rich in its interior than on the surface. What could be going on?

The answer may have come from studies of nearby stars which are almost identical to our Sun in terms of size and age – so-called 'solar twins'. Compared with most of these stars, the solar surface has a lower proportion of elements that vaporize (or condense) at high temperatures, such as calcium and aluminium. These refractory elements are, however, present in much higher proportions in meteorites and in the rocky planets of the inner Solar System. The pattern can be explained if the refractory elements that might have gone into the atmosphere of the young Sun instead went into the formation of rocky, Earth-type planets – so-called terrestrial planets. And evidence in favour of this idea comes from studies of the solar twins. Solar twins that have giant planets in close orbits, where they would have acted to inhibit the formation of terrestrial planets, do not show the pattern of refractories that the Sun does. But some of the solar twins that do not have giant planets in close orbits do have a depletion of refractories! It is tempting to conclude that these stars, like the Sun, have families of rocky planets. Depletion in refractories seems to be the signature of the presence of terrestrial planets. The good news is that this suggests that there are other planets superficially similar to the Earth.

But this is not entirely good news. Only about 10 per cent of the solar twins fit this category. If only stars with similar proportions of refractories to the Sun have families of terrestrial planets, that means a further reduction in the number of possible locations for technological civilizations. The number was already down to 0.6 per cent of all the stars in the Galaxy; now, it is down to 0.06 per cent. That is, just 6 stars out of every 10,000 in the Milky Way. I won't try to quantify any further reductions in this number, but it is worth bearing in mind that each additional feature that makes our place in the Universe special is operating to winnow out the candidates even further from this already tiny proportion. There are plenty of those additional features that make our Solar System special. But before moving on to consider them, having described the birth of the Sun it seems a pity not to complete the story of its life and death, even if that is not directly relevant to the chances of finding other intelligences in the Galaxy.

UNTIL THE SUN DIES

Many popular accounts, and even some textbooks, simply mention that the Sun will swell up as it ages to become a red giant star that will engulf the inner planets of the Solar System, including the Earth. This is true up to a point; but the full story is much more subtle, and much more interesting, than these simple accounts suggest.

Stars like the Sun maintain their output of energy by converting hydrogen into helium in their hearts. The helium builds up as a kind of stellar ash, so as time passes a central core of helium develops, with the hydrogen 'burning' going on in a growing core composed of a mixture of hydrogen and helium. This is why the output of heat from the Sun has slowly increased in the 4.5 billion years since it formed, gradually pushing the habitable zone outwards. But there comes a time when there is not enough hydrogen available in the core to maintain this process, and without the heat from nuclear fusion available to hold it up, the core collapses. The Sun was initially 30 per cent helium throughout; today, the Sun's core consists of 65 per cent helium, and 35 per cent hydrogen. In about 5 billion years' time no hydrogen will remain at the centre, and the core will collapse. This makes the core hotter, as gravitational energy is converted into heat. The core settles into a so-called 'degenerate' state, which you can think of as being like a solid ball of helium. At the same time, hydrogen from farther out in the star falls on to the core and begins to 'burn' in a shell around the core. The overall result is that more heat is flowing out from the heart of the star, and this makes what is left of the outer layers swell up enormously, so that the star becomes a red giant. During this phase of its life, a star occupies a region of the Hertzprung-Russell Diagram known, logically enough, as the red giant branch. But it doesn't stay there.

As the hydrogen-burning shell deposits more helium 'ash' on the core the core shrinks due to the extra weight of material. As the core shrinks it becomes hotter. Eventually the helium nuclei are squeezed so much they begin to combine together to form carbon. The degenerate core behaves more like a solid than a gas, so heat from this 'helium burning' spreads rapidly through the core leading to a chain reaction known as the helium flash. Within just a few seconds the core explodes. The energy released lifts the outer layers and blows some of the outer atmosphere of the star away into space, reducing the pressure so that the core is no longer degenerate. All of this stabilizes the core once again, but at a higher temperature than when it was held up by hydrogen burning. The star then settles down into a more compact state once again, with helium burning going on in the core and a shell of hydrogen burning surrounding the core. But helium burning does not provide as much energy as hydrogen burning, and for a star which starts out with the same mass as our Sun the helium fuel is exhausted within about 150 million years.

For a star with the same mass as our Sun, helium burning is the end of the line. After the helium fuel in the centre of the star is exhausted, the core shrinks again, forming a degenerate carbon core surrounded by shells of helium burning and hydrogen burning. The outer layers of the star swell up once more, but the conditions in the heart of the star never become extreme enough to allow further phases of nuclear burning. When the supply of hydrogen and helium is exhausted, the star becomes a ball of material with about half as much mass as the

Sun has today, packed into a solid lump roughly the size of the Earth. Then it cools off, and the remaining tenuous outer layers settle down on to what has become a white dwarf star.

As far as the fate of the Earth is concerned, the key factors are just how big the Sun will get during each of its two phases of life as a giant star, and just how much mass it will lose along the way. If no allowance is made for the mass loss, the calculations tell us that the first time it swells up to become a red giant the Sun will swallow the inner planets Mercury and Venus, while the second time it expands it will engulf Earth and Mars. In other words, the maximum radius of a giant star with the same amount of mass that the Sun has today is bigger than the radius of the orbit of Mars. This is why some books tell you that the Earth will one day be swallowed by the Sun.

But mass loss makes the situation more complicated, and more interesting. Astronomers at the University of Sussex have worked out the implications in detail. Because a star like the Sun has already lost about 20 per cent of its mass by the time it becomes a red giant, its gravitational hold on its atmosphere will be weaker, and this means it will actually expand slightly more than if it had not lost any mass, reaching out to a radius of 168 million km, compared with the Earth's present-day orbital radius of 150 million km. By then, however, the effect of the weakening of the Sun's gravitational grip will have allowed the Earth's orbit to expand out to a distance of 185 million km from the Sun! By the time it reaches the peak of the second phase of its expansion, about a hundred million years after its first experience as a red giant, the Sun will have lost so much mass – about 30 per cent of the mass it has today – that it only reaches a maximum radius of 172 million km, while the Earth will have moved out to 220 million km. This is very close to the present-day orbit of Mars.

Or rather, the Earth would drift out that far, if it still existed. Unfortunately, there are still other factors to consider – drag, and tides. As the Sun expands, the outer layers of its atmosphere will stretch out past the Earth, and although this region, known as the chromosphere, is very tenuous, it will be enough to drag on the Earth and send it spiralling slowly in towards the Sun. At the same time, as the Earth orbits the swelling Sun it will raise tides on the surface of the Sun that will drain energy from the Earth's orbital motion, holding the planet back in its orbit and also making the radius of the orbit shrink. The combined effect is that the Earth will indeed be swallowed by the Sun, in about 7.6 billion years from now, roughly half a million years before the Sun reaches its maximum size as a red giant for the first time and long before it fades away to become a cooling white dwarf.

POSTPONING DOOMSDAY

As far as the prospects for life on Earth are concerned, however, this is all largely irrelevant. The Sun has been gradually increasing in brightness throughout its life, and yet the temperature at the surface of the Earth has been roughly constant for all that time, in the region where liquid water can exist. The explanation is that the concentration of greenhouse gases in the atmosphere, notably carbon dioxide, has slowly been declining, so that a decreasing greenhouse effect has balanced the increasing solar warmth. Researchers such as Jim Lovelock have pointed out that this process has gone just about as far as it can – if there were no human impacts on the atmosphere, all the carbon dioxide would be gone in the near future. Lovelock calculates that with the Earth warming in line with the increasing heat from the Sun, within a hundred million years it will flip into a hot state more extreme than anything life on our planet has experienced so far. The most conservative estimate is that the Earth will be uninhabitable within a billion years. The true situation is far worse than this, because at present we are adding carbon dioxide and other greenhouse gases to the atmosphere and making the planet warm faster; but I shall ignore that for now.

On an astronomical timescale, the first thing that will happen as the Earth warms is that more water will evaporate from the oceans. Since water vapour is a greenhouse gas, this will speed up the warming, until the oceans boil away. But that won't happen for billions of years. So if our technological civilization survives, or if a technological civilization on another planet somewhere else in the Galaxy is faced with a similar problem, there is ample time for a technofix. One possible solution to the problem is disarmingly simple, and involves only slightly more advanced technology than we already have. It involves boosting the Earth in its orbit to move it away from the Sun as the Sun heats up.

The trick is to take energy from a passing asteroid and give it to the Earth – not once, but repeatedly, at intervals of about 6,000 years. Of course, the asteroid would have to be just the right size and in just the right orbit to do the trick, which is where the technology comes in. A spaceprobe could be sent out to the Kuiper Belt, a region of the Solar System beyond the giant planets, densely populated by chunks of ice and rock. There, it could be used to nudge a chunk of material about a hundred kilometres across into the appropriate orbit around the Sun, sending it diving close by the Earth. In that close encounter, energy from the orbital motion of the asteroid would give the Earth a boost, expanding its orbit by a tiny amount. The asteroid would lose energy and its orbit would shrink, but that doesn't matter.

What does matter is that the same thing would have to be done every 6,000 years, for hundreds of millions of years. It might be possible to put the asteroid in a particular orbit so that after it had zipped in past the Earth and back out from the Sun it looped around one of the giant planets, retrieving its lost energy (at the expense of the giant planet, whose orbit would shrink by a minute amount) and setting it up for another pass by the Earth. Or it might be possible to send a von Neumann machine to sit quietly out in the Kuiper Belt and do the necessary every 6,000 years. That would ensure that, even if civilization collapsed, the job of nudging the Earth outwards would continue. But it would still involve a chunk of material five times bigger in diameter than the one which killed off the dinosaurs passing within 15,000 km of the Earth every 6,000 years. That doesn't leave much room for error. You'd certainly need to have great faith in your von Neumann machine to set such a long-term project going.

In principle, then, the habitability of the Earth could be extended into the far future, perhaps for another 6 billion years, until the Sun dies. But the hazards involved in this kind of technofix highlight another unusual feature of our place in the Universe. In spite of what happened to the dinosaurs, our Solar System seems to be a rather orderly place, with a neat arrangement of planets in tidy circular orbits, and any leftover bits and pieces of cosmic rubble pretty much confined to the Kuiper Belt and another belt of debris, the Asteroid Belt between Mars and Jupiter. Did it have to turn out that way? Probably not, but the fact that it did is one of the reasons why we are here to ask such questions.

4
What's So Special about the Solar System?

The first thing that's special about the Solar System is that it exists – in particular, that the rocky planets like the Earth exist. Planets, especially rocky planets, are the most complicated, and therefore interesting, things that astronomers have traditionally studied. The only entities in the Universe that are more complicated are living things, and astronomers are now turning their attention to life itself, within the discipline of astrobiology. Things that are much smaller than a human being, such as atoms and simple molecules of substances such as oxygen and water, are only capable of a limited range of interactions, because of their simplicity. A living animal is a complex entity capable of a vast variety of interactions with its environment, because it is made of complex molecules such as DNA and proteins.

A living planet, like the Earth, is even more complex, providing the stage for a complicated interaction of astronomical, geological, chemical and biological processes which work together to maintain the whole system; this is the basis of James Lovelock's idea of the Earth as a single living entity, which he calls Gaia. A planet like Venus or Mars may lack the biological contribution, but it is still the site of astronomical, geological and chemical processes. But by the time we get to an object the size of a star (or even, to some extent, a giant planet like Jupiter) things have become more simple again, because the gravitational force inside such an object is so great, and the temperature is so high, that complex molecules, and even atoms, are destroyed, and the only interactions are ones involving atomic nuclei and particles such as protons, neutrons and electrons. Planets exist on the edge between two extremes of simplicity; and they exist on the edge in another sense as well.

Recent studies made by the infrared telescope of the Spitzer space observatory have shown that clusters like the one in which the Sun formed can be dangerous places for the survival prospects of the dusty circumstellar discs in which rocky planets form. The dangers cannot be completely quantified; but they are none the less real.

With a few tiny exceptions, all of the material in the Solar System is rotating in the same direction. The Sun itself rotates on its axis once in just over 25 days, the planets and asteroids go round the Sun in the same direction that the Sun is rotating, and the various moons also orbit their own planets in this preferred direction. This tells us that the Sun, planets and moons formed together from a single cloud of material that started out with some random overall rotation. As the cloud shrank, it had to spin faster (like the spinning ice skater pulling in their arms), conserving the property called angular momentum. Angular momentum depends on the mass of an object, how fast it is spinning, and how spread out it is. In order for the Sun to shrink down to its present size, it had to lose angular momentum. Some was carried away by streams of matter being ejected from the young Solar System, like water spraying out from a garden sprinkler; but some was stored in a dusty disc surrounding the young Sun but stretching far out into space. This is where the planets formed. A planet like Jupiter, far from the centre of the Solar System, has a lot of angular momentum even though it has much less mass than the Sun. The Sun got most of the mass, but the planets got most of the angular momentum.

But not all planetary systems operate in such a tidy way. In many systems that contain hot jupiters, the planets are orbiting in the opposite direction to the rotation of their parent star. These so-called retrograde orbits have probably been produced by gravitational interactions between the planets and a distant companion star, in a wide orbit around the parent star. These interactions would tilt and elongate the orbits of Jupiter-like planets, shifting them into highly elliptical orbits in which they would lose energy every time they passed close by the parent star and be forced into a 'wrong way' circular orbit. This neatly accounts for the presence of Jupiter-like planets orbiting close to their parent stars; it also means that any Earth-like planet in such systems would have been destroyed in the process. So these systems are of no interest in our search for the reason why we are here, except to confirm that only solitary stars like the Sun are likely to provide homes for life forms like us.

Discs like the one in which the planets of our Solar System formed can be seen around some young stars, confirming that our ideas about how the Solar System formed are correct. They are known as protoplanetary discs, or PPDs, and some of them extend out more than a thousand times farther from their parent star than the Earth is from the Sun – more than 30 times farther than Neptune, the outermost major planet of our Solar System, is from the Sun. These discs are still losing matter (and angular momentum) as it is blown away into space. In many cases, the inner regions of the discs are distorted by the gravitational influence of planets forming within them. But such discs are quite fragile, by astronomical standards, and can easily be destroyed before planets – at least, planets like the Earth – get a chance to form. The problems arise if the young star forms in too close proximity to a hot, massive star, which is all too likely in the kind of clusters where stars like the Sun form.

TOO HOT TO HANDLE

The most massive stars on the Main Sequence, the O-type stars, produce a lot of ultraviolet radiation, which heats up any gas and dust in a disc around the star and blows it away into space. This is one reason, as well as the short lifetime of an O star, that the chances of finding intelligent life on a planet orbiting one of these stars are negligible. But the radiation from an O star is so powerful that it can disrupt the discs around any smaller stars in its vicinity. Until recently, it was hard to quantify just how far the danger zone around an O star extends. But the Spitzer observatory was able to identify PPDs around stars in the Rosette Nebula, an active star-forming region about 5,200 light years away from us, in the direction of the constellation Monoceros.

In other regions of the Milky Way, far from any O stars, just under half of all young stars with masses in the range from one tenth of to five times the mass of the Sun have dusty discs around them. The Spitzer team looked for such discs around a thousand stars in the Rosette Nebula, at various distances from O stars. They found the same proportion of PPDs around stars at least 1.6 light years from any O star, but only half as many for stars closer than this to an O star, with fewer and fewer discs the closer they looked to the O stars. For comparison, the nearest star to the Sun today, Proxima Centauri, is just over 4 light years from us. The implication is that any young star which passes within about a light year of an O star is likely to have its dusty disc evaporated away, within about a hundred thousand years. This would destroy any prospect of rocky planets like the Earth forming. But if giant planets like Jupiter, or like the 'super jupiters' seen orbiting some stars, had already formed before the close encounter, they would not be destroyed by the ultraviolet radiation. They would still be there long after the O stars had burnt out and the danger had passed. This is a sobering reminder that the detection of super jupiters does not necessarily imply the existence of other earths.

The Sun was lucky to have been born in a relatively small, relatively diffuse cluster where this kind of close encounter with an O star did not occur. Just how small and diffuse that cluster was can be inferred from the remarkable circularity of all the orbits of the planets and other objects of the Solar System, out to Neptune and, indeed, beyond.

For the moment, take it as read that the planets of the Solar System did form in a dusty disc, in almost circular orbits – the explanation of how they formed follows shortly. During the early stages of this process, the stars in the young cluster were close enough together that it would have been quite easy for one of them to pass by another within a distance less than the distance our Sun is now from the planet Neptune. This must have happened to some of the planet-forming systems in the cluster, thoroughly disrupting the orbits of the planets. Indeed, in our Solar System the orbits of many of the comets beyond about 50 times the distance from the Earth to the Sun (50 astronomical units) are markedly elliptical, and inclined at angles to the plane in which the planets orbit. Something has stirred them up. These objects are too far out from the Sun to be disturbed by the gravity of the planets – even by the gravity of Jupiter – so they must have been stirred up by the gravity of a star passing close enough to affect them but not close enough to affect the orbits of the planets. That implies a passing distance of about a thousand AU, but the circularity of the planetary orbits rules out the possibility that the Sun has experienced an encounter with a star coming closer than 100 AU since the Solar System formed.

Allowing for the frequency of stellar encounters in the cluster where the Sun formed, which depends on the density of the cluster, and for the time it takes for the cluster to dissolve away, these figures imply that the cluster was less than three light years across, and with a relatively low density of stars – only about 3,500 stars in the whole cluster, judging by the radioactive evidence mentioned in the previous chapter. The combination of exploding neighbours, O stars and close encounters makes anything much larger or more crowded than this an unlikely place for an orderly planetary system like our Solar System to have formed. This further highlights the special nature of our place in the Milky Way. To see just how special the Solar System is, though, we first need to know a bit more about the layout of the planets and other objects that orbit the Sun.

THE GEOGRAPHY OF THE SOLAR SYSTEM

Apart from the Sun itself, the Solar System today is made up of four components – rocky planets like the Earth, smaller pieces of rocky debris known as asteroids, the giant gaseous planets, and icy pieces of debris that include dwarf planets such as Pluto and the progenitors of comets.

Mercury, the planet closest to the Sun, orbits at an average distance of only 0.39 AU, but it has an elliptical orbit that takes it as close as 46 million km to the Sun and as far out as 70 million km. It takes 88 Earth days to orbit the Sun once, but it spins on its axis only once every 58.6 of our days, locked in a gravitational grip that makes each rotation of Mercury take exactly two thirds as long as its year. Because it is so close to the Sun, on the side of the planet facing the Sun the temperature exceeds 450 °C; but because there is no atmosphere on Mercury to carry heat from one side of the planet to the other, during the long night the temperature on the dark side falls to −180 °C. With a diameter of 4,880 km, Mercury is only 50 per cent bigger than our Moon, and like our Moon it is smothered in craters that are evidence of a phase of intense bombardment when the Solar System was young. The largest crater, the Caloris Basin, is 1,340 km across, and must have been made by the impact of an object about 150 km across. Planets like Mercury are definitely not likely places to find technological civilizations.

For a long time, the planet Venus, next out from the Sun, was thought of as a possible home for life. In terms of size, Venus is very nearly a twin of the Earth, with a diameter of 12,100 km, only 650 km less than that of the Earth itself. It is covered by a thick layer of cloud that makes it impossible for optical telescopes to view the surface of the planet, and this encouraged speculation that the clouds might conceal oceans and continents like those of the Earth, perhaps covered in tropical life. Alas for this romantic view of Venus, space-probes have now landed on its surface, where the pressure of the thick, carbon dioxide atmosphere is 90 times the pressure of the air at sea level on Earth, and thanks to an intense greenhouse effect the temperature exceeds 500 °C. Radar mapping of the surface shows that it has highlands and lowlands, but it is all bare rock, with no trace of water or life.

There is nothing odd about the orbit of Venus, which is very nearly circular, with a radius of 0.72 AU (108 million km), so that Venus takes 225 of our days to orbit the Sun once. But there is something odd about the rotation of the planet itself – it is the only planet in the Solar System that rotates from east to west, backwards compared with the rotation of the Earth. It does so very slowly, taking 243 Earth days to turn on its axis once. The most likely explanation is that Venus received a massive blow from space some time in its history, reversing the spin that it must have inherited from the formation of the Solar System. And although there is clearly no life on Venus, such an impact may have had dramatic consequences for life on Earth, which are described in Chapter 6.

Earth, of course, orbits the Sun at a distance of 1 AU (149,597,870 km) once every 365.242199 days. The average temperature at the surface of the Earth is about 15 °C, comfortably in the range where liquid water can exist, and there is plenty of oxygen in the air, which is a key factor in allowing energetic life forms like ourselves to exist.

Mars, the next planet out from the Sun, is tantalizingly close to providing a home for life forms like ourselves. About 50 per cent farther out from the Sun than we are, Mars has a slightly elliptical orbit taking it as close as 1.38 AU to the Sun and as far out as 1.67 AU. It is only just over half as big as the Earth, with a diameter of 6,790 km, and has only a little more than one tenth of the mass of the Earth. Partly because of its low mass, it has only been able to hold on to a tenuous atmosphere, which is mostly carbon dioxide; the pressure at the surface of Mars is the same as the pressure in our atmosphere 35 km above sea level.

Coincidentally, the length of a day on Mars is almost the same as the length of a day on Earth – 24 hours 37 minutes, compared with our 23 hours 56 minutes. But Mars takes 687 Earth days to orbit the Sun once. Temperatures on Mars today range from about −26 °C to about −110 °C, too cold for liquid water to flow; but there is evidence that liquid water did flow on Mars in the past, carving channels in the surface. Almost certainly this is because Mars had a thicker atmosphere when it was young, providing a greenhouse effect strong enough to keep the temperature above freezing in spite of its distance from the Sun. The fate of this atmosphere provides another important clue to the scarcity of intelligent, technological civilizations in the Milky way. It wasn't lost solely because of the weak gravity of Mars. The planet also lacks a strong magnetic field, and this meant that there was nothing to shield it from electrically charged particles that stream outwards from the Sun – the solar wind – which scoured the atmosphere away over billions of years. The Earth's strong magnetic field seems to have been a major factor in providing us with a planet where conditions suitable for life lasted long enough for intelligence to evolve.

Beyond the orbit of Mars there is a region of the Solar System filled with pieces of rubble left over from the formation of the planets. This forms a thin disc, known as the Asteroid Belt, which stretches from about 1.7 AU to about 4 AU. It contains more than a million asteroids each at least a kilometre across, and uncounted numbers of smaller pieces, down to the size of grains of sand. But there are only ten asteroids bigger than 250 km across. The largest of these is Ceres, which has a diameter of 933 km and contains more than a quarter of all the mass in the Asteroid Belt; it orbits the Sun once every 4.6 Earth years, at a distance of 2.8 AU. Together with the two next-largest asteroids, Vesta and Pallas, Ceres may represent the kind of building block from which planets formed. A small amount of the lesser cosmic rubble is in elliptical orbits that take the fragments of rock closer to the Sun than the Earth, crossing the Earth's orbit on the way. If the Earth happens to be in the way, they enter our atmosphere at high speed. Small particles burn up in the atmosphere as meteors; larger pieces of rock can reach the ground, as meteorites.

There is not enough mass in all the asteroids put together to make a planet even the size of Mercury, but fragments of rock from the Asteroid Belt which fall to the Earth as meteorites show that much of this material was part of a large object (or objects) early in the life of the Solar System, and this got broken up during the processes which formed the four remaining rocky planets. This is an important clue to the way the planets formed. Another important clue comes from the location of the Asteroid Belt, which depends on the gravitational influence of the next planet out from the Sun, the gas giant Jupiter.

Jupiter is so big that it influences the entire structure of the Solar System, and is a key to understanding the reason why we are here. Its mass is 0.1 per cent of the mass of the Sun, which doesn't sound impressive by stellar standards but corresponds to 317 times the mass of the Earth, and is more than twice the mass of all the other planets in the Solar System put together. With a diameter eleven times that of the Earth, Jupiter has a volume more than 1,300 times bigger than that of our home planet. It orbits the Sun once every 11.86 of our years at a distance of 5.2 AU, and the closest it ever gets to Earth is 590 million km.

Most of the mass of Jupiter is in the form of hydrogen and, to a lesser extent, helium, so its composition resembles that of a failed star rather than a rocky planet like the Earth. It also resembles a star like the Sun in another way – because of its strong gravitational pull, it has a retinue of smaller objects orbiting around it, in the same way that the planets orbit the Sun. The four largest of these moons were discovered by Galileo, and 400 years ago the evidence that these moons orbit around Jupiter, not the Earth, helped to persuade people that the Earth is not the centre of the Universe.

Beyond Jupiter there are three more giant planets, each fascinating in its own right but only peripherally important to the story of life on Earth. Saturn, famous for its spectacular ring system, orbits the Sun once every 29.46 Earth years at a distance of 9.5 AU and has a diameter of 120,536 km, just over nine times that of the Earth, with a mass 95 times that of the Earth. Uranus, the next planet out from the Sun, has a slightly elliptical orbit, taking it as close as 18.3 AU to the Sun and as far out as 20.1 AU. It takes 84 of our years to complete one orbit. The mass of Uranus is 8.7 times the mass of the Earth, and with a diameter only about four times that of the Earth it is small for a gas giant. Some astronomers prefer the term 'ice giant' for Uranus and the last major planet in the Solar System, Neptune, which orbits the Sun at a distance of 30 AU. Neptune, which takes 165 Earth years to complete one orbit, has a diameter 3.8 times that of the Earth (similar to Uranus) but a mass 17 times that of the Earth, twice as much as Uranus.

Beyond Neptune there is a frozen counterpart to the Asteroid Belt, the region known as the Kuiper Belt, which extends from about 30 AU to about 50 AU from the Sun. It is made up of material left over from the formation of the Solar System, and because it is so far from the Sun much of this material is in the form of ice – not just water ice, but things like frozen ammonia and frozen methane. Statistical analysis of the orbits of observed Kuiper Belt Objects, or KBOs, suggests that there are at least 70,000 of them with diameters bigger than a kilometre, and that around half of these have diameters larger than 100 km. The largest of these objects are called dwarf planets; they include Pluto, with a diameter of 2,300 km, which used to be regarded as a planet in its own right, but is now known to be smaller than some of the other objects in the Kuiper Belt. But the total mass of all the objects in the Kuiper Belt put together is only about 40 times the mass of the Earth, less than half the mass of Saturn.

Beyond even the Kuiper Belt, we come to the edge of the Solar System. Analysis of the orbits of comets shows that many of them come from far out in space, where, in order to account for the observations, there must be a spherical shell, or cloud, around the Solar System, at a distance of 50,000 to 100,000 AU (0.75 to 1.5 light years!) from the Sun. There, about as much mass as there is in the Kuiper Belt is shared out among several thousand billion separate objects, tens of millions of kilometres apart from one another, drifting around the Sun in their lonely circular orbits. This is known as the Oort Cloud. The objects in the Oort Cloud are about a third of the way to the nearest star, and although just about held in the Sun's tenuous gravitational grip, from time to time interactions with passing stars or interstellar gas clouds can send some of these icy objects falling in towards the Sun, where they heat up, releasing streaming clouds of gas and becoming comets.

MAKING PLANETS

Armed with this understanding of the geography of the Solar System, we can begin to understand how the planets formed, and why the Solar System is special. Working back inwards from the Oort Cloud, the key to this understanding is the way the giant planets formed, and the role of Jupiter in particular.

The giant planets cannot have been formed simply by sticking smaller pieces of material together to build up gas (or ice) giants. A lump of rock with several times the mass of the Earth, in the orbit of Jupiter or beyond, could indeed attract gas from a cloud like the one in which the Sun formed and build up to the size of the giants – but it would take longer than the present age of the Solar System to make planets like Uranus and Neptune this way in their present orbits, because there was little gas available that far from the Sun. The only alternative is that all four of the giant planets formed closer to the Sun than Saturn is today, but farther out than the present orbit of Jupiter, where there was plenty of gas but it was also cold enough for icy material to exist. They would have formed close to one another, in a cloud of gas swarming with planetisimals made of ice and rock. It was just by chance that four giants emerged from the mêlée. Although inevitably one of the four had to be bigger than the others and came to dominate the orbital dynamics of the system through its gravitational influence, there was nothing inevitable about the way that Jupiter became so much bigger than the other planets put together; it is just one more of those things that distinguishes our Solar System from other planetary systems.

Computer simulations show that in these circumstances the orbit of the largest planet (Jupiter) moves slowly inwards towards the Sun, while the orbits of the three smaller planets (Saturn, Uranus, Neptune) move slowly outwards, conserving angular momentum. At first, this process proceeds smoothly. But about 700 million years after the Solar System formed, it passes through a so-called resonance, a situation in which Saturn takes exactly twice as long to orbit the Sun as Jupiter. In this situation the combined influence of Saturn and Jupiter produces a powerful, rhythmic influence on the smaller planets in the outer Solar System. The simulations tell us that as a result Uranus and Neptune both suddenly moved farther out from the Sun, with the orbit of Neptune doubling in size and sending the planet into the inner region of what was then a much larger Kuiper Belt. These disturbances shook up the smaller objects in the outer Solar System, sending many of them falling in towards the Sun, where a large number smashed into the inner planets, many outwards into the depths of space, and some close to the giants where they were captured and became moons. It was only after this interval of turmoil that the giant planets settled into fairly stable orbits close to their present ones – although it turns out that no orbit in the Solar System is, in fact, truly stable.

But the simulations also show that well-ordered planetary systems like our own are rare. When researchers from Northwestern University in the United States ran dozens of simulations starting out with a disc of gas and dust around a star like the Sun and ending up with planets, they found that in most cases the end product was 'full of violence and drama', with hot jupiters being common and the planets usually in elliptical and unstable orbits. 'These other planetary systems don't look like the Solar System at all ... Conditions must be just right for [a system like] the Solar System to emerge.' The 'vast majority' of planetary systems are, they say, 'very different' from our own.

In our Solar System, though, the intense bombardment of the inner Solar System during the rearrangement of the giant planets' orbits essentially marked the end of the formation of the rocky planets, including the Earth. The beginning of the formation of the rocky planets occurred when the cloud of leftover material from the formation of the Sun settled into a disc around the young star. Most of the material in the cloud, like the material of the Sun itself, was in the form of hydrogen and helium. But there was a trace of dust, no more than 2 per cent of the original material, in the form of particles as fine as the particles of cigarette smoke. Heat from the young Sun blew much of the gas away, but the rotation of the original cloud ensured that the dust settled into a disc around the young Sun – a proto-planetary disc like the ones seen around young stars today.

Within the disc, all the particles were moving in the same direction around the Sun, like runners going round a track. This meant that when they bumped into one another, they did so relatively gently, not in head-on collisions, giving the particles a chance to stick to one another. The tendency to stick may have been helped by electric forces produced by particles rubbing against one another, in the same way that you can make a child's balloon stick to the ceiling after rubbing it on a woollen sweater. Another important factor was turbulence in the gas, creating swirling structures like whirlwinds which gathered pieces of material together and gave them a chance to interact. Computer simulations show how objects as big as Ceres can form in this way – provided the particles can stick together.

Something else may also have helped the particles to stick together – something else that is special about the Solar System. Studies of pieces of rock from meteorites show that the dusty disc around the young Sun contained tiny globules of material, known as chondrules, formed by melting at temperatures between 1,200 °C and 1,600 °C. Molten, or partly molten, blobs would be more sticky and encourage the buildup of larger lumps of stuff in the disc. But how did they get so hot? The most likely explanation is that the heat was released by radioactive elements that had been sprayed by the death throes of a nearby star into the gas cloud from which the planets formed. One possibility, matching the evidence discussed in Chapter 3, is that a supernova exploded close to the cloud that became the Sun just before the Sun formed; it is even possible that the blast wave from this explosion triggered the collapse of the gas cloud that became the Sun and Solar System. Supporting evidence for this idea comes from measurements of the proportions of various isotopes found in meteorites. Radioactive aluminium-26 seems to have been present in the proto-Solar System from the beginning, but a pulse of iron-60 arrived about a million years later. This matches what we know about the fate of a very large star, with more than 30 times as much mass as the Sun. In the late stages of its life, the star first blows away much of the outer layers of material, which by then is relatively rich in aluminium-26, in a wind easily strong enough to cause any nearby gas cloud to collapse. The star only explodes at the very end of its life, showering the neighbourhood with elements including iron-60.

There is a rival idea, developed in Barcelona by Josep Trigo-Rodriguez and colleagues, which suggests that the radioactive material was fed into the Solar System as it was forming from a much less massive star which came much closer to the Sun. The right proportion of isotopes could have come in the wind of material being blown away from a star with only six times as much mass as our Sun in the last stages of its life. But the star would have to be very close to the Sun to do the trick – closer than 10 light years – which makes such an event unlikely, statistically speaking.

The important point is that both scenarios are unlikely, but something of the kind seems to have been necessary for the formation of rocky planets. The implication is that while other systems may contain giant planets and small pieces of rocky debris forming extended asteroid belts, systems like ours, with a few giant planets and a few Earth-sized planets, may be rare.

The good news, though, is that once objects a kilometre or so in size have built up, which seems to have happened within about

100,000 years of the formation of the Sun, the rest of the planet-building process, though messy, is easy to explain.

MAKING THE SOLAR SYSTEM

Douglas Lin, of the University of California, Santa Cruz, has looked in detail at what happens to the solid chunks of material that build up in the disc around a young star like the Sun. One crucial feature of his calculations is the way the solid objects interact with the gas that is still present in the disc during the early stages of planet formation. Because of the interaction between pressure, gravity and rotation, the gas at any chosen distance from the central star moves around the star slightly more slowly than the speed with which the particles and lumps of material are moving in their orbits. This means that the particles are overtaking the gas; in effect, as Lin puts it, 'running into a headwind that slows them down and causes them to spiral inward, toward the star'. A piece of material a metre across can halve its distance from the star in this way in just a thousand years, and the bigger the pieces grow the faster they move inwards – up to a point.

That point is picturesquely dubbed the 'snowline'. It is the distance from the star where frozen water, ammonia and other volatile substances evaporate; in the case of the Sun at a distance of between 2 AU and 4 AU, between the orbits of Mars and Jupiter. This is why the boundary between the rocky planets and the icy objects in our Solar System lies where it does.

At the snowline, water vapour released by the icy grains as they evaporate changes the properties of the gas in such a way that it now rotates faster than the solid grains, giving them a boost which tends to make them move outwards in their orbits. So material piles up at the snowline, where grains are packed closer together and can quickly grow into larger lumps. Within a million years of the formation of the Sun, many of these lumps are a kilometre or so across and very little dust remains. As they grow, and as gas is being dissipated from the inner part of the disc by the heat of the Sun, the planetisimals, as they now are, are less influenced by interactions with the gas, and many of them migrate inwards towards the Sun, into the region where rocky planets are found today. The exact positions that the planets end up in when this migration stops depends on many factors, including the temperature in different regions of the disc and the size of the planet, but the overall picture is clear from many computer simulations.

Planetisimals gather up the remaining dust gravitationally and collide and merge with one another, with the survivors settling into roughly circular orbits which have been swept clean of debris. Chunks of material left over from these collisions may still be with us, in the form of some of the asteroids. Because there is more dust to feed on farther out from the Sun, embryonic planets grow bigger farther out. According to Lin's calculations, at a distance of 1 AU from the Sun a planetary embryo can grow to one tenth of the mass of the Earth within 100,000 years, but then all the available dust is gone; at a distance of 5 AU, there is more dust and an embryo can continue to grow for a few million years, reaching a size of about four Earth masses. But this isn't the end of the story. Interestingly, Lin points out that there is no room for any more planets in our Solar System today – the planets we have are as close together as the complex interaction of gravitational forces between them will allow. It is very likely that more planets formed when the Solar System was young, but that the surplus were ejected from unstable orbits before the present stable pattern was established.

It cannot be a coincidence that Jupiter, the largest planet in our Solar System, lies just beyond the snowline; but astronomers are still not able to explain just how a Jupiter-sized planet ended up in a stable orbit there. Interactions between an embryonic planet in the outer part of the Solar System and the gas in the disc, still significant that far from the Sun, explain why the embryonic Jupiter ended up close to the snowline, and accumulated a great deal of gas from the material available there. But what stopped it spiralling inwards into an orbit like those of the many 'hot jupiters' that have now been discovered? If it had done so, it would have pushed any rocky planets in the inner Solar System into the Sun ahead of it.

Once Jupiter had formed, it helped the other giant planets to form by stopping the inward flow of material in the disc and by disturbing the orbit of planetisimals so that many of them migrated to the outer part of the Solar System. The first effect aided the formation of the second gas giant, Saturn; the second effect provided enough frozen chunks to make the massive cores of the ice giants, Uranus and Neptune. All of this, prior to the processes which sent the giant planets into their present orbits and disturbed the Kuiper Belt, only took about 10 million years after the formation of the Sun. But the formation of the Earth took a lot longer.

MAKING THE EARTH

Computer simulations show that by about a million years after the collapse of the cloud from which the Sun and planets formed, there would have been twenty or thirty objects in the region between the Sun and the present orbit of Mars, ranging from about the size of the Moon (roughly 27 per cent of the diameter of the Earth today) to about the size of Mars (roughly 53 per cent of the diameter of the Earth today). They would have been accompanied by a huge number of smaller planetisimals, which would have been swept up by the larger objects in a series of collisions, while the larger objects themselves collided with one another and merged until eventually only four or five large objects were left – the objects that became Mercury, Venus, Earth and Mars, plus at least one other Mars-sized object. The heat of the young Sun would have destroyed complex molecules and driven gases outwards, so that these four (plus one) proto-planets would have been mostly made of iron and silicates, plus stable compounds of carbon.

But although Mercury and Venus have no moons at all, and Mars has only two tiny moons which are irregularly shaped pieces of rock captured from the Asteroid Belt by the planet's gravitational field, the Earth has a moon which is the largest, in proportion to its parent planet, of any moon of any of the eight major planets in the Solar System. To an astronomer, the similarity in sizes is so close that the Earth–Moon system is more properly regarded as a double planet. So how did such an unusual system form?

The most likely explanation is that the Earth began life as a near-identical twin to Venus, with a thick rocky crust, while another planetary object, about the size of Mars, formed nearby. The most likely place for this object to form would have been at one of two places known as Lagrangian points. These lie 60 degrees ahead or behind the Earth but in the same orbit around the Sun. They are places where the combined effect of the gravitational pull of the Sun and the gravitational pull of the Earth is to produce a kind of gravitational pothole, a place where small objects can accumulate and stick around for a long time. The Lagrangian points are used today as stable parking places for satellites, such as the Herschel infrared telescope, which need to be kept far enough away from the Earth not to suffer interference from natural or man-made radiation from our planet. A small object which is not quite at the exact Lagrangian point wobbles slightly to and fro about the point itself, like a swinging pendulum; the orbits of artificial satellites at these points have to be adjusted from time to time, using their rocket motors, to keep them in place. But if a large natural object grew up out of cosmic debris near to one of the Lagrangian points of the Earth's orbit these oscillations would get bigger and bigger, soon becoming so extreme that the object would bash into the Earth itself. This would have happened within 50 million years of the formation of the original crust of the Earth.

The name of the hypothesized proto-planet is Theia, after the Greek goddess who gave birth to Selene, the Moon goddess. Theia formed with the other planets of our Solar System about 4.6 billion years ago. Theia's orbit became unstable when its mass exceeded a critical value, leading to the collision which formed the Earth–Moon double planet about 4.53 billion years ago, roughly 30–50 million years after the other rocky planets had formed.

Such a collision would not be like two pieces of solid rock colliding and chipping pieces from one another. Astronomers refer to this collision as the 'Big Splash', and the image that conjures up accurately indicates what happened when the Earth was young and was struck a glancing blow by an object the size of Mars. So much energy of motion would have been released by the collision that the incoming object would have been completely destroyed, and the entire surface of the Earth itself would have melted. The outer layers of the incoming object would also have melted, and mixed with the molten material from the Earth's surface, with much of it being flung off to make a ring of debris around the planet. Meanwhile, the dense, metallic core of the incoming object would have sunk through this molten outer layer and been absorbed into the core of the young Earth. The lighter material from the incoming object and from the Earth's original surface splattered out into space in this way would have contained about ten times the present mass of the Moon; most of it escaped entirely into independent orbits around the Sun, becoming asteroids, but some was captured in a ring of material around the Earth. As the surface of the Earth cooled and formed a new, thinner crust, the material in this ring coalesced into the Moon, repeating in miniature, but far more quickly, the process by which the planets themselves formed around the Sun. Computer simulations suggest that about 2 per cent of the original mass of Theia ended up in the ring of debris, and about half of this fused together to form the Moon. The time taken to complete the formation of the Moon would have been only about a month.

Other objects created out of the ring of debris may have got stuck in Lagrangian point orbits for as long as a hundred million years, before the gravitational influence of other planets shook them out of these gravitational potholes, allowing many of them to crash into the Earth or the Moon.

Evidence to support this model of how the Earth–Moon system formed comes from samples of rock brought back from the Moon. These show that it has exactly the same composition as the Earth's crust. And seismic measurements of moonquakes made by instruments left on the lunar surface show that it has no significant metallic core; the radius of the core is certainly less than 25 per cent of the radius of the Moon, whereas the radius of the Earth's core is about 50 per cent of the radius of the planet. The Moon's core contributes only a few per cent of the Moon's total mass, but the Earth's core makes up nearly a third of the planet's mass. Because of the lack of iron, the overall density of the Moon is much less than the density of the Earth. Earth has a mean density of 5.5 grams per cubic centimetre, but the Moon has a density of only 3.3 grams per cubic centimetre.

The age of Moon rocks even gives us a precise date for when this dramatic event happened – 4.4 billion years ago, almost as soon as the Sun had formed. There's also more circumstantial evidence; such a glancing blow explains why the Earth rotates so rapidly, once every 24 hours, while moonless Venus rotates only once every 243 of our days. The glancing blow which formed the Moon would actually have set the Earth spinning even faster, so that it would have had a day some five hours long after the impact, and it has been slowing down ever since. The off-centre impact also gave the Earth its tilt, which is the reason why we have seasons, but the presence of such a large moon orbiting the Earth has since acted as a gravitational stabilizer, stopping the tilt from varying very much over geological time. Incidentally, a combination of the extra iron in the Earth's core and the rapid spin probably explains why our planet has a strong magnetic field. All of these influences may, as we shall see, have been crucial in allowing the emergence of a technological civilization on Earth.

There is one more persuasive piece of evidence that collisions like this did happen when the Solar System was young. Spaceprobes flying past Mercury have measured the strength of its gravitational pull and found that in spite of its small size it has a relatively high density. The Moon resembles the crust of the Earth without a core, but Mercury resembles the core of the Earth without a crust. The natural explanation is that a much larger object originally formed in the orbit of Mercury, but that early in the life of the Solar System it was hit, not in a glancing blow but a head-on collision, by another proto-planet. In a head-on collision, all the lighter material would have been blasted away into space, leaving only the heavy core behind.

The impact model explains why only one out of eight planets in the Solar System has a moon comparable in size to the parent planet; but it also implies that such double planets are rare.

THE SPECIAL ONE

Overall, our Solar System seems to be special – or to be, at least, a rare kind of planetary system – because the orbits of the planets are nearly circular and far enough apart that they do not have a big influence on one another today. In most of the other planetary systems that are known, the orbits of the giant planets are more elliptical. (This finding is not just the result of giant planets in elliptical orbits being easier to find: it is statistically significant.) Such systems are also more likely on dynamical grounds; it is easy to explain how planets come to be in such orbits, but harder to explain how they can be in circular orbits. Nobody yet knows why some planetary systems, including our own, have settled into a more regular state. It may be just statistical chance: if a rare configuration is possible then if there are enough planetary systems it must occur somewhere. Or it may be related to the amount of gas there was in the disc around the young Sun, dragging on the planets as they grew. That in turn would be linked to the environment in which the Solar System formed and the presence of material dumped on it by a nearby supernova or a giant star. For whatever reason, the more diverse astronomers find planetary systems to be, the more special the Solar System seems.

The latest category of planetary systems to be discovered highlights this point. Several nearby stars are now known to be accompanied by planets larger than Earth but smaller than Neptune, orbiting so close to their parent star that they zip round once every few weeks – clearly not likely locations for other earths. But there is some hope for astronomers seeking other systems like our own; the star 23 Lib has a planet with roughly the same mass as Jupiter moving in a nearly circular orbit that takes it once around the star every fourteen years, just a couple of years longer than it takes Jupiter to go round the Sun. Perhaps it also has a family of rocky planets too small to have been detected yet.

Even in a nearly circular orbit, the influence of Jupiter on the formation of the rocky planets has not been entirely benign. The Asteroid Belt represents the remaining rubble from a planet that might have been, if the available material had not been swallowed up by Jupiter itself or deflected by Jupiter's gravitational influence away from the region entirely. Mars, the nearest rocky planet to the Asteroid Belt, is only half the size of the Earth, with about one tenth of the mass of the Earth. If it had been as big as the Earth, it might have had a strong enough gravitational pull and other properties necessary to hold on to a thick atmosphere, keeping it warm by the greenhouse effect. Venus, on the other side of the Earth and away from the immediate proximity of Jupiter, indeed has roughly the same size and mass as our home planet, although there the greenhouse effect has gone to extremes. Without Jupiter, Mars too could probably have grown to the size of the Earth – and if the missing 'asteroid planet' had done the same, our Solar System might have started out with three Earth-like planets that were the right distance from the Sun for liquid water to flow – always assuming, and it is a big assumption, that such planets could form at all without the presence of Jupiter. On the other hand, if Jupiter had been a little bigger, or the Earth had formed a little closer to Jupiter, the same processes that restricted the growth of Mars would have limited the size of the Earth.

But however many, or few, Earth-like planets formed, could their orbits have remained stable without the influence of Jupiter on the Solar System as a whole? The evidence suggests not. It is actually impossible to calculate precisely how the orbits of all the planets change over long intervals of time, because of the way each planet simultaneously influences all of the others, if only by a tiny amount. If there were just one planet going round the Sun, its orbit, a simple ellipse, could be calculated perfectly for all time. But adding just one more planet creates a 'three body problem' that is impossible to solve precisely. The only solution is to compute the orbits step by step, allowing one planet to move on a little bit, calculating the resulting influence on the other planet and moving it on a little bit, then calculating how that change affects the first planet, and so on.

Modern computers are so powerful that it is possible to make calculations of this kind involving all the major planets of the Solar System to see how the orbits change over billions of years. But such computations show that the orbits of the inner planets are affected by mathematical chaos. This means that in some cases a very small change in the starting conditions of the calculation leads to a huge change in the evolution of the orbits. It is an example of what is sometimes called the 'butterfly effect', from the notion that the flap of the wings of a butterfly in Brazil could, in principle, affect the track of a hurricane in the Caribbean.

In the case of the Solar System, the computer calculations show that the orbits of the four giant planets are very stable, and not prone to chaos. But only just – if either Jupiter or Saturn had been a little more massive, or if the two gas giant planets had been slightly closer together, or if there had been a third gas giant of comparable size to them in the outer Solar System, chaos would have ensued. Worse, when simulations of the orbits of the inner planets are run thousands of times with slightly different changes in the starting parameters, in one computation in every hundred chaos occurs. Because of gravitational interactions with the other planets, in particular Jupiter, the orbit of Mercury, the smallest of the rocky planets, becomes stretched so much that Mercury can pass close by or even collide with Venus, disturbing the entire inner Solar System so much that Venus itself is, in some of these scenarios, flung out of its orbit and collides with the Earth. The fact that this occurs in 1 per cent of the simulations means that there is a one in a hundred chance that this will happen in our Solar System some time before the Sun dies. It's not anything to lose sleep over, but it shows that the stability of planetary systems cannot be taken for granted. Indeed, by studying the strange orbits of three planets discovered orbiting the star Upsilon Andromedae, and in effect running their calculations backwards in time, astronomers have inferred that the system used to harbour a fourth planet, which has been ejected entirely from the system as a result of just this kind of orbital chaos.

The inference, once again, is that our Solar System is not typical but is a rare example of a family of planets with more or less stable orbits. Just how rare such a system is, though, it is impossible to say.

What is certain is that within that pattern of stability, overall the influence of Jupiter has been beneficial, in terms of making the Earth a suitable place for a civilization like ours to arise. It was involved in the rearrangement of the outer Solar System which led to the intense bombardment of the inner planets, including the Earth–Moon double planet, shortly after the rocky planets formed. Many of the impacts during this heavy bombardment involved lumps of rock more than a hundred kilometres across, and such impacts today would be disastrous for life on Earth. But by clearing out much of the debris left over from the formation of the planets in this way, the Late Heavy Bombardment reduced the amount of cosmic rubble around for such collisions in the future, while parking most of the remaining debris in the Asteroid Belt, the Kuiper Belt and the Oort Cloud. Although objects can still escape from these special niches and plunge sunward, ever since the Late Heavy Bombardment Jupiter has acted as a kind of sentinel, protecting the inner Solar System from most of the material that falls in from farther out. Many of these objects are deflected by Jupiter's large gravitational pull, either being shifted into orbits that prevent them crossing the Asteroid Belt or being swallowed up entirely by Jupiter, as in the case of the comet Shoemaker-Levy 9, which broke apart and collided with Jupiter in July 1994; this was the first direct observation of such a collision of Solar System objects.

Geological evidence tells us that, throughout most of Earth history, on average our planet has been hit by an object at least 10 km across once every hundred million years. It was an impact of this size that contributed to the 'death of the dinosaurs', 65 million years ago. But if Jupiter had not been around to first clear out most of the original Solar System debris and then guard us from objects like Shoemaker-Levy 9, such impacts would have occurred every 10,000 years, not every hundred million years. It is hard to see how animal life could have evolved at all under such circumstances, let alone evolved intelligence and developed a technological civilization.

The impacts which battered the early Earth, though, may have been an essential prerequisite for our existence. How did a once molten planet (actually, twice molten if our ideas about the origin of the Moon are correct) hang on to enough water to fill the oceans? The answer is that it didn't – the water came later. But where did the water come from?

Until recently, it was a profound puzzle that water could survive in liquid or vapour form in the disc of material from which the planets formed. Water molecules ought to be broken up by the ultraviolet radiation from a young star, in much the same way that oxygen molecules are today broken up, high in the Earth's atmosphere, by ultraviolet radiation from the Sun. But the water must have been there, or it wouldn't be here today; and during the first decade of the twenty-first century the spectral signatures of both water itself and the hydroxyl radical (OH; a kind of water molecule with one hydrogen atom removed) were seen in the planet-forming discs around many young stars. In some cases, there is even indirect evidence for the presence of icy, comet-like objects. The explanation for the survival of water in these discs came in 2009, from researchers at the University of Michigan. They discovered that water vapour shields itself from the damaging effects of ultraviolet radiation, because the incoming radiation is in effect used up by being absorbed by the water molecules on the outside (the side nearest the star) of a cloud. As fast as the water molecules are split apart by the radiation to release hydrogen and OH, they re-form through ordinary chemical reactions and have to be broken apart again.

Because there is always plenty of hydrogen around in the clouds from which stars form, the overall result is that all of the available oxygen is converted into water. The water-rich inner part of the cloud is protected, in a way very similar to the way the ozone layer of the atmosphere protects life at the surface of the Earth from solar ultraviolet radiation. In the stratosphere of our planet, reactions involving ultraviolet radiation are constantly converting oxygen molecules into ozone, but other chemical reactions convert the ozone back into oxygen as fast as it is formed. The overall effect is that there is a permanent layer of ozone high over our heads, but very little solar ultraviolet gets through to the ground. Within the clouds of material in which planetesimals start to form, there is a similar shielding effect, with very little ultraviolet radiation penetrating the outer layer. This allows complex molecules to exist, and even to become more complex as they interact with one another, providing a rich source of organic compounds with which planets can be seeded, even within a few astronomical units from the parent star.

But how does the water, and the complex organic molecules, get down to a planet like the Earth, after the planet has cooled off sufficiently to stop the water boiling away? It comes from planetesimals which always contained water, and, being small, never got hot enough for the water to vaporize. To put this in perspective, today the Earth contains less than 1 per cent water and less than 0.1 per cent carbon, but the debris in the Asteroid Belt contains 20 per cent water (combined with other substances) and a bit less than 5 per cent carbon (more about the carbon in the next chapter). If the Earth had formed out of the same kind of material as the asteroids are made of, just beyond the orbit of Mars, it would have had an ocean hundreds of kilometres deep and an atmosphere composed of a thick layer of carbon dioxide, raising the surface temperature through the greenhouse effect so high that the water would have boiled away, strengthening the greenhouse effect further and making it too hot for life. Even if the Earth had only ended up with twice as much water as it actually has, it would have been almost entirely covered by water, with very little land, if any, on which our kind of life could evolve. Too much water is as bad as too little when it comes to developing a technological civilization. This is another example of Earth being 'just right', balanced between one extreme and another.

In our Solar System, as well as its many other roles in making the Earth a suitable home for life, Jupiter is in just the right place to have sent our way just enough asteroids rich in water when the Earth was young. And not just asteroids – the Late Heavy Bombardment, which at first sight looks like a catastrophe for the Earth, also involved icy material from the outer part of the Solar System, comet-like objects rich in water and organic molecules. Supporting evidence for this scenario has come from studies of the isotopes of krypton and xenon in the Earth's mantle, the layer just beneath the crust. These match the proportions found in meteorite samples and confirm that the Earth acquired its volatile materials from impacts at a late stage of the formation of the planet. In fact, the impact of just one moderately sized object from the 'water belt' could have provided enough water to fill the Earth's oceans. The recent discovery of water locked up in Moon rocks also matches this model.

If Jupiter had formed farther out from the Sun, computer simulations show that it would still have been possible for an Earth-sized planet to have formed, but only about five astronomical units from the Sun. It would have had plenty of water, but all of it locked away as ice. Without a Jupiter at all, there would still have been plenty of

WHAT'S SO SPECIAL ABOUT THE SOLAR SYSTEM?

water, but all locked up in asteroids, which never got a chance to form an Earth-sized planet at all.

In our special Solar System, though, the Earth did form, along with a large moon, and it does have oceans of liquid water. That was enough for life to get started on the planet. But there is much more to the story of how that life evolved intelligence and developed a technological civilization.

5
What's So Special about the Earth?

The Earth is a rocky planet. That may not look so special in terms of the Solar System – four out of the eight major planets are made of rock. But in the Universe at large, rocky planets are much less common than this suggests. It all depends on the relative proportions of carbon and oxygen, which are the two most common elements after hydrogen and helium, in the disc of material around a star in which planets form.

LIKE A DIAMOND IN THE SKY

Rocks are made of silicate materials, which involve combinations of silicon atoms with oxygen atoms – essentially, oxides of silicon, although sometimes combined with other elements. If there is plenty of spare oxygen around in the region where a planet forms, virtually all of the silicon will get locked up in silicate and you will have a rocky planet. But if there is no spare oxygen there will be no silicate and no rocky planets. The alternative is to have plenty of carbon, because the proportion of carbon and oxygen produced by stellar nucleosynthesis is nearly in balance, with slightly more carbon in some stars and interstellar clouds, and slightly more oxygen in others. In the Galaxy at large there is roughly twice as much oxygen as carbon, and ten times more carbon than silicon, but carbon dominates over oxygen around some stars. Carbon and oxygen have a great affinity for each other, chemically speaking. Whichever is the least abundant in the cloud from which a new planetary system forms will get locked up in compounds like carbon monoxide and carbon dioxide, with the surplus of the more abundant atoms available to take part in reactions with other elements.

This is not just theory. Observations made at infrared wavelengths have shown that many stars at a late stage of their life are surrounded by clouds of material, including dust grains rich in carbon compounds, being blown away into space; by far the most common of these is silicon carbide. There are also more complex carbon compounds, and as these get distributed through space they may play a significant part in the emergence of life on suitable planets at a later date. Tellingly, though, out of all the stars now known to harbour planets, most have a higher carbon-to-oxygen ratio than the Sun has.

There is actually very little carbon on Earth, compared with things like silicate rock. In our Solar System, most of the carbon in the inner region where the rocky planets formed seems to have been blown away as carbon monoxide gas when the Sun was young, and got incorporated into the structure of asteroids and comets, farther out in the Solar System. Like water, carbon was brought down to Earth later on, by the impact of these objects, after the planet had solidified – another reason to be grateful to Jupiter and the Late Heavy Bombardment.

But in planetary systems where carbon dominates over oxygen, this will not happen. Solid carbon compounds such as silicon carbide, and even solid carbon itself in the form of grains of graphite, will be the building blocks of planets, even planets in orbits like the orbit of the Earth around the Sun. An Earth-sized planet that formed in this way would have a crust of graphite, but deep below the surface the pressure would be strong enough to produce layers of silicon carbide and of crystalline diamond, giving a whole new depth of meaning to the rhyme 'Twinkle, twinkle, little star'. On the surface of such a planet, the modest amount of oxygen available would be in the form of carbon monoxide, with more carbon present in the form of methane (a compound of carbon and hydrogen). Under some circumstances, such a planet would have lakes and oceans of tar.

Just how much this reduces the chance of finding planets like ours is an open question, although clearly this is not good news for SETI enthusiasts. And like many of the factors that make the Earth a suitable home for life, it is a question of time as well as of place. Regions of the Galaxy that contain a higher proportion of heavy elements ('metals') have higher carbon-to-oxygen ratios. This means that regions of the Galaxy much closer to the centre of the Milky Way than we are are unlikely to have rocky planets like the Earth, quite apart from the other reasons why such regions are inhospitable to life forms like us, and that a wave of what you might call 'carbonification' is sweeping outwards through the Galaxy as time passes. None of these factors can yet be properly quantified. But it is clear that all of this reduces the likelihood of the existence of Earth-like worlds even farther. And even within the Solar System, the nature of the Earth's silicate crust seems to be both unique and vital for the emergence of life forms like us.

A PLANETARY JIGSAW PUZZLE

The crust of the Earth is divided up into several large 'plates' and a few smaller ones, which fit together like the pieces in a spherical jigsaw puzzle. But unlike jigsaw pieces, these plates are in constant motion, leading to what is still sometimes called continental drift, although the term plate tectonics (meaning 'plate building') is preferred today. Many lines of evidence, including magnetic surveys of the sea floor, seismology, and direct observations from space of the way the continents are moving today, combined in the second half of the twentieth century to produce the modern understanding of plate tectonics. It explains how and why volcanoes and earthquakes happen where they do, why there are sometimes Ice Ages but sometimes the Earth is ice free, and even why species diversify and go extinct.

At the heart of this understanding of plate tectonics is the discovery of sea-floor spreading. There are cracks in the sea floor, notably along a line running roughly north–south through the Atlantic Ocean, where molten material from beneath the crust (magma) wells up to the surface in a ridge, then pushes out on either side of the crack, where it sets. The result is to produce a kind of ocean-floor conveyor belt, which gradually, but steadily, pushes the continents on either side of the ocean apart. In the case of the North Atlantic, this process is widening the distance between Europe and America by a couple of centimetres per year – about the rate at which your fingernails grow. And this widening of the ocean has been measured directly using instruments on orbiting satellites.

If this was all there was to plate tectonics, the Earth would have to be growing steadily larger to accommodate all the new sea floor being laid down at ocean ridges. But in fact sea floor is being destroyed as rapidly as it is being made. The destruction happens in regions where sea floor gets pushed under a continent, diving down back into the interior of the Earth via a deep trench. This doesn't happen everywhere that sea floor meets a continent – it doesn't happen around the edges of the Atlantic Ocean, for example. But the archetypal example of such a destructive region is down the western coast of the Pacific Ocean, where Pacific sea floor being pushed westward dives under the land mass of the Eurasian continental plate. It is no coincidence that this entire region, including Japan, is prone to earthquakes and volcanic activity. The Pacific Ocean is shrinking at about the same rate as the Atlantic Ocean is expanding, and the destruction of sea floor is a violent process. The creation of sea floor is also a violent process, of course; but it mostly happens far out to sea, where the earthquakes and underwater volcanoes do not bother us. An exception is Iceland, where a section of the mid-Atlantic ridge has pushed up above sea level.

Overall, the process is simply convection. Hot, fluid material from beneath the surface of the Earth rises, spreads out as it cools, then dives back down into the interior. Simple. Of course, there are complications. Hot spots beneath the surface can break through continents and crack them apart to form new oceans (this seems to be happening in East Africa today). And sometimes continents collide, as the sea floor between them shrinks away to nothing, throwing up new mountain ranges, such as the Himalayas. But the underlying principles are clear. It is also clear that these processes can only operate on a planet, like the Earth, with a relatively thin crust of solid material on top of the fluid layers beneath.

The Earth's crust is not the same thickness everywhere, however. The continental crust ranges in thickness from about 35 km to 70 km, whereas the crust beneath the oceans is more uniform and averages about 7 km thick. To put this in perspective, the diameter of the Earth is a bit less than 13,000 km. On that scale, the entire crust is no more significant than the skin of an apple is compared with the whole apple.

Even so, the differences between oceanic and continental crust are important – perhaps, literally vitally important. As well as being thicker than oceanic crust, continental crust is also less dense. The density of sea-floor crust is about 3 grams per cubic centimetre, but the density of continental crust is only about 2.7 grams per cubic centimetre. This is partly a reflection of the fact that oceanic crust is essentially a solid slab of uniform material, while continental crust is made up of various bits and pieces that have been jumbled up and stuck together by the processes of plate tectonics. But it explains why it is oceanic crust, not continental crust, that is destroyed at the deep trenches where material descends back into the Earth's interior. The less dense material naturally rides on top of the more dense material, encouraging it on its way down into the depths. But there is something else that encourages the process – water.

The regions where oceanic crust is being destroyed in deep trenches along the edges of continents are called subduction zones (there are also places where one piece of sea floor is sliding under another piece of sea floor, but that is a detail we can ignore). The sea floor which is carried down into the depths of a subduction zone is modified by interactions with water, which permeates through cracks in the rock and can get hot enough to boil, encouraging chemical reactions which change the composition of the rock. At the edge of a continent the rock itself is also covered in a layer of sediment, rich in the remains of organic life, which has been washed down from the continent. As this mixture dives below the continental crust, it gets hotter and it is squeezed by extreme pressures – at a depth of 50 km, the pressure is 15,000 times the atmospheric pressure at the surface of the Earth, and the temperature reaches hundreds of degrees Celsius. One effect of all this is to squeeze water out of the rocks and into the material that surrounds the sinking slab of rock. There, it encourages the rock to melt, in the same way that spreading salt on ice encourages it to melt. Blobs of magma grow near the sinking slab, and because molten magma is lighter than solid magma, these blobs of molten material gradually rise upwards, penetrating the crust and creating a chain of volcanoes above the region where sea floor is being destroyed. Without water, none of this would happen; without water, there would be no plate tectonics.

It is easy to see how important all of this is for the existence of life on Earth – especially our kind of life. While this process is going on, gases bubble out of the magma and make their way to the surface, where they are released at volcanic vents. The most important of these gases are water vapour, carbon dioxide and nitrogen. The nitrogen largely comes from the organic remains mixed in with the sediment carried downwards with the sinking slab of lithosphere. All of these gases are important to living things on the surface of the Earth, so life on Earth is intimately linked in to the cycles involving the non-living rocks. One of the most important aspects of this is that the combination of living and non-living processes helps to maintain the balance in the atmosphere of greenhouse gases, notably carbon dioxide, regulating the Earth's temperature through a kind of global thermostat that maintains conditions suitable for life.

There are other aspects of this volcanic activity that are important for the emergence of a technological civilization. As the molten rock rises through the crust it cools, and some of it solidifies on its way to the top. One of the ways to quantify what happens is in terms of the amount of silica in different rocks. Silica is just another name for silicon dioxide, and it is present in virtually all rocks. Basalt released by volcanic activity is by far the most important component of oceanic crust. It contains about 50 per cent silica (in terms of its weight), but the crust that forms the continents has a different composition, including 60 per cent silica – another factor in making continental crust lighter than oceanic crust. Some rocks, of course, contain even more silica than the average – granite, for example, is about 75 per cent silica. The reason why continental crust contains relatively more silica than basalt is that other things have been taken away from the mixture typical of basalt on its journey to the surface from a subduction zone. Those other things include minerals such as quartz, which crystallize out of the molten magma, leaving it relatively richer in silica as it continues its rise, and compounds rich in metals such as iron, copper, silver and gold, which solidify out higher up. This is why precious metals are found, for example, in the young mountains of South America – wealth which brought the Spanish Conquistadors to the continent in search of El Dorado, and bankrolled the great days of the Spanish Empire. And it was this kind of process, operating long ago when the continents were arranged in patterns very different from those of today, that laid down the European deposits of metals such as copper and iron which were so essential for the industrial revolution. It is the combination of a thin crust and water that has made technology possible on Earth, leaving aside the reasons why intelligent life evolved on Earth. Without the metals, intelligence alone could not have produced a technological civilization capable of signalling to, and perhaps travelling to, the stars. But what made the continents where these processes took place and on which our technological civilization developed?

CREATING CONTINENTS

One of the key insights provided by plate tectonics is that ocean basins are temporary features of the globe, but the continents are permanent, although they may get torn apart and welded back together in different configurations as the eons pass. They also grow (which means that as time passes the area of our planet covered by sea is steadily declining). Volcanic activity along the continental edges associated with subduction zones adds new material to the continents, and when continents collide oceanic crust can get crumpled up and become part of mountain ranges. Of course, continents are also being constantly worn away by wind and weather, with sediments being washed down to the sea bed. But the rate at which new material is being added to the continents far outweighs the erosion. If we go back far enough in time, there can have been no continents at all.

This is obvious when we think about how the Earth formed. After the impact which gave birth to the Moon, our planet settled down as a molten blob of material, forming a nearly spherical shape (only 'nearly' spherical because its spin would have made the equator bulge a little). It would have set into an almost featureless sphere. If the Earth were smoothed out into such a sphere today, there would be enough water in the oceans to form a single shallow sea covering the entire globe. The average depth of the oceans is 3,800 m, more than four times the average height of the continents, and two thirds of the planet is covered by sea; so if the solid surface of the Earth were smoothed out, it would be covered by water to a depth of just under 3 km.

If all that water had fallen as rain onto the cooling surface of an undisturbed young Earth, it might well have remained a water world, with all that that implies for life and technology. But a great deal of the water – perhaps all of it – was brought down to Earth by impacts from space, and it is now thought that later impacts are what promoted the growth of continents and triggered the processes of plate tectonics.

Even below the oceans of the young Earth, there would have been cracks in the crust, caused by magma welling up from the interior. Embryonic plates would have been jostling one another as they were carried around by convection currents, and some must have got pushed under others, creating mini subduction zones and pushing up the first volcanic islands. When one platelet pushes under another, bits get scraped off, and these rock scrapings would have contributed to the growing mini continents. But there is evidence that rather more than half of the continental crust that exists today had already formed by 2.5 billion years ago, and this implies a much more dramatic burst of continental growth early in the history of our planet than anything that could be produced by the jostling of embryonic plates. The explanation seems to be that the Earth was struck by a series of major impacts between about 3.8 billion years ago (the end of the Late Heavy Bombardment) and 2.5 billion years ago. As a result, more than half (perhaps as much as two thirds) of the Earth's continental crust was produced in just 700,000 years, less than a fifth of its lifetime to date.

Evidence for the impacts is clear enough. There is direct evidence for huge impacts in the structure of geological features known as cratons, found in some ancient rocks. The most striking examples come from a region in southeast Africa and a region in northwest Australia. They are so similar that they were clearly once part of a large land mass that got split apart. And there is indirect evidence from the battered face of the Moon, from which astronomers can estimate how many impacts of different sizes affected the Earth and its neighbour during different intervals of geological time. The conclusion is that between 3.8 billion years ago and 2.5 billion years ago the Earth may have been struck nine or ten times by objects with sizes in the range from 20 km to 50 km in diameter.

To put this in perspective, the asteroid implicated in the death of the dinosaurs some 65 million years ago was only about 10 km across. An object bigger than 20 km in diameter, arriving at a speed of about 20 km per second, would scarcely even notice the existence of a thin layer of water, only 3 km deep, around the Earth. The appropriate image is not tossing a pebble into the sea, but dropping a brick into a puddle. The incoming piece of debris would smash straight into the underlying crust, generating so much heat that it would melt the surface into a lake of liquid rock perhaps 500 km across.

But how could such an event give a boost to continent building? Andrew Glikson, of the Australian National University, has a plausible explanation. At such a distance in time, any explanation must be speculative to some extent; but his idea brings together all the pieces of the puzzle seamlessly, and is backed up by calculations of the effect of such an impact made by Jay Melosh, of Purdue University. A primordial version of plate tectonics, involving only thin pieces of crust, could continue relatively quietly, with hot molten material rising in plumes, pushing the thin plates apart as it spread out at the surface, then cooling and sinking back down into the interior of the Earth. But if a large asteroid struck just above a rising mantle plume, the molten lake of rock that it produced would be hotter than the rising plume, so the plume would be diverted. The molten lake would extend over and (crucially) under the edges of the young plates. Instead of reaching the surface through a crack in the crust, the plume would become a hot column of magma rising under the already existing plate. There, it would produce violent volcanic activity, with molten rock breaking through the crust and building up on it to form volcanic mountains, increasing the thickness of the crust and stimulating the more vigorous form of plate tectonics that we know today. Depending on the exact geometry of the situation, a lesser plume, split off from the main one, could break through to the surface on the other side of the molten lake as the lake cooled and solidified, making a new chain of volcanic islands.

Attractive though this scenario is, it is by no means certain that it is the reason why plate tectonics and continent building got a big boost early in the history of the Earth. But what is certain is that the combination of a thin crust and plenty of water is essential for plate tectonics to take place at all in the way that we know it. The thin crust is a legacy of the impact that created the Moon, and another legacy of that impact is the dense, iron-rich core of the Earth, which also turns out to be essential for the development of our kind of civilization.

A FIELD OF FORCE

In a certain kind of science fiction story, spaceships and people are often surrounded by almost magical 'force fields' that protect them from attackers. It's a nice idea, but not very practical on the scale of spaceships and people (even assuming such fields exist) because of the enormous amount of energy that would be required to produce a shield of this kind. But the whole Earth, and in particular life on the surface of the Earth, is indeed protected from certain kinds of danger from space by exactly this kind of force field, generated by swirling currents of molten metal deep in the interior of the planet. It is the Earth's magnetic field, or magnetosphere, and although it cannot shield us from incoming asteroids, it does protect us from dangerous charged particles from space, known as cosmic rays.

Working out what lies beneath our feet is a harder task, in some ways, than working out what the stars are made of. Stars may be far away, but we can analyse light from stars, and use spectroscopy to determine which hot atoms are making that light. But seismologists have made great progress in understanding the inner structure of the Earth, by studying the way vibrations associated with earthquakes spread around the globe. These waves travel at different speeds in different kinds of rock, and some of them travel deep below the surface, passing through the interior of the planet before being detected on the other side of the world. They also get bent and reflected, in the same way that light waves get bent and reflected by something like a block of glass. With enough observations it is possible to build up a picture of the interior structure, reminiscent of the way it is possible to build up a picture of the inner structure of the human body using X-rays.

Working downwards from the surface, the combined oceanic and continental crust of the Earth makes up only 0.6 per cent of the volume of our planet, and only 0.4 per cent of its mass. The layer below the crust, which is called the mantle, extends down to a depth of about 2,900 km and makes up 82 per cent of the volume of the Earth. The mantle itself is divided into two regions with slightly different properties, the upper and lower mantles, but the difference is not important as far as generating the Earth's magnetic field is concerned. That happens even deeper, in the core of the Earth. The inner core is a solid lump, thought to be mostly iron and nickel, with a diameter of about 2,400 km (a radius of 1,200 km). So the top of the inner core is about 5,200 km below the surface of the Earth. But all the action we are interested in here occurs in the outer core, which is a liquid layer, rich in iron and nickel, that extends from the top of the inner core to the base of the mantle, a span of 2,300 km. Overall, the core makes up only 17.4 per cent of the volume of the Earth (it is about the same size as the planet Mars), but it is so dense, squeezed by the weight of all the material above, that it contains a third of the mass of the Earth.

Seismic studies show that the outer core is liquid, and laboratory experiments tell us that under the pressure that exists there, the temperature at which an iron-nickel mixture liquefies is about 5,000 °C. So that must be the temperature (only slightly less than the temperature at the surface of the Sun) of the liquid material in which the swirling currents responsible for generating the Earth's magnetic field occur.

How can the deep interior of the Earth still be so hot, more than 4 billion years after the planet formed? It's partly because the outer layers of the planet provide good insulation, like a blanket keeping the heat in. But it is also because the core is laced with radioactive elements, such as uranium and thorium, which are still releasing heat as they decay, even after all this time. And the core is laced with radioactive elements because of the environment in which the Solar System formed, from a cloud of material that had been enriched by debris from a nearby supernova or the winds from a giant star. This is another hint that civilizations like ours on planets like ours may not be too common, since even Earth-sized planets may lack molten cores, and therefore lack protecting magnetic force fields. In about 4 billion years from now, the entire core of the Earth will have solidified, and it will lose its magnetic field.

The magnetic field is a result of physical currents of electrically conducting metal swirling around in the outer core and acting as a dynamo, producing electric currents which in turn generate magnetic fields. The region occupied by the magnetic field around the Earth, the magnetosphere, is actually shaped like a teardrop, because it is squashed in by a wind of charged particles from the Sun on one side, but stretches off into space on the other. On the side of the Earth facing the Sun, the boundary between the magnetosphere and the solar wind of particles, the magnetopause, lies about 10 Earth radii (more than 60,000 km) above the surface of our planet; on the other side it stretches roughly as far as the distance to the Moon, beyond 60 Earth radii. Electrically charged particles in the solar wind, things like protons, travel at speeds of several hundred kilometres per second most of the time, with bursts travelling at about 1,500 km per second when the Sun experiences bouts of activity known as solar storms. The Earth, and the entire Solar System, is also bombarded by particles from deep space known as cosmic rays.

All of these particles could do severe damage to life if they reached the surface of the Earth – they are essentially the same as the particles produced by radioactivity or in nuclear explosions. But because they are electrically charged, they are funnelled by the Earth's magnetic field towards the poles, where they interact with molecules of gas high in the atmosphere to produce the colourful activity of the auroras.

Even so, during solar storms the electrical activity caused by the arrival of these particles at the Earth can disrupt communications and distort the magnetic field locally to such an extent that power lines can be affected and blackouts can be caused in high-latitude countries such as Canada. An increase in intensity of the strength of solar-wind particles can also knock out satellites, including communications satellites, and pose a health hazard to any astronauts unlucky enough to be in space at the time. So how bad would the total removal of the magnetic field be? As it turns out, we know just how bad it can be – because it has happened, and more than once.

As you might expect for a magnetic field produced by swirling currents of molten metal, the Earth's magnetic field is not steady. It varies in strength from time to time, and the exact location of the magnetic poles drifts across the surface of the Earth. A record of past magnetism is preserved in rocks that were being laid down at different times – as the molten rock sets, magnetic field gets trapped in it, so that the rock today preserves, like a fossil, an indication of both the strength and the direction of the magnetic field that existed long ago. From such evidence, the way the field has changed can be reconstructed by geologists and compared with the fossil evidence of what life was like at the time.

For reasons that are not understood, from time to time the magnetic field gradually dies away completely to nothing then builds up again, either in the same configuration as before or with the magnetic poles reversed, so that what was the north magnetic pole becomes the south magnetic pole, and vice versa. The fossil record shows that when the magnetic field dies away, many species of life on the surface of the planet go extinct. The obvious explanation is that land-dwelling species in particular are killed off by radiation from space which reaches the Earth's surface during magnetic reversals. In fact, though, it doesn't matter what the exact connection is. What matters is that there is a link between the absence of the magnetic field and death on the surface of our planet. Clearly, the existence of a protecting magnetosphere is an important factor in allowing life forms like us to have evolved on Earth.

A reversal typically takes several thousand years to complete. Once the field has become established in a certain orientation it may stay that way for as little as 100,000 years or as long as tens of millions of years. Sometimes, there are even bursts of reversals, with four or five flips taking place in a million years or so, and these tend to be associated with more extreme extinction events. If you want to worry about such things, there is evidence that the Earth's magnetic field is getting weaker at present, and has been for thousands of years; if the trend continues it will disappear quite soon, by geological standards.

So another reason why we are here is that the Earth has a strong magnetic field, and the reason it has a strong magnetic field is that it has a large metallic core, formed as a result of the impact in which the Moon was formed. We have, indeed, a lot to be thankful to the Moon for. And this is by no means the end of it. Before looking at the ongoing role of the Moon in maintaining the Earth as a suitable home for civilization, though, it's worth looking at how Venus and Mars have suffered through a lack of the geological features that make Earth special.

VENUS AND MARS

As we have seen, Venus is nearly the twin of the Earth in size, and might be expected to have similar geological activity. But that is not the case. A partial explanation is that the thick atmosphere of Venus, rich in carbon dioxide, has induced a runaway greenhouse effect on the planet, leaving it with no water to, among other things, lubricate the processes of plate tectonics. But that is not the whole story; Venus doesn't seem to have experienced plate tectonics as we know it here on Earth at all. Venus has now been thoroughly mapped by radar from orbiting spaceprobes, so we can 'see' its surface in great detail. Although that surface does have highlands and lowlands, with evidence of volcanic activity where molten material from the interior breaks through the crust, there is no evidence of sideways motion – there is no Venusian continental drift.

One curious feature of the surface of Venus is that it is not as heavily cratered as the surfaces of the Moon, Mercury and Mars. Studies of the surfaces of those other bodies, especially the Moon, have revealed the rate at which impacts have continued to occur since the Late Heavy Bombardment. The evidence comes from across the inner Solar System, all the way from Mars to Mercury, so Venus must have experienced the same kind of long-term bombardment. But there is no trace of the Late Heavy Bombardment itself on Venus, or of anything in the 3 billion years that followed, just a more modest amount of cratering which corresponds to the number of impacts Venus is thought to have received over about the past 700 million years. The best explanation that anyone has been able to come up with for this is that about 700 million years ago molten rock from the interior of Venus broke through the crust in a catastrophic global event and spread over the entire planet before it cooled and set into a featureless surface, obliterating traces of previous impacts and providing a blank slate on which meteorites could begin to make their marks again.

This explanation works if Venus has a thick crust – as thick as the Earth's crust would have been if it were not for the impact that created the Moon. The thick, insulating crust seals in the hot interior of the planet, where radioactivity makes the temperature rise until a critical point is reached and the whole surface is cracked and flooded with magma. With the heat released, the planet settles down, the surface solidifies, and the whole process starts again. Venus may have been resurfaced in this way several times since the formation of the Solar System. This kind of resurfacing never happens on the same scale on Earth, because heat is escaping steadily from the interior at the spreading zones where magma rises to the surface.

Supporting evidence for the idea comes from what might seem at first an unrelated fact – Venus has no significant magnetic field. That means it lacks a large, metal-rich core. The Earth's large metallic core and thin crust both result from the impact that created the Moon; without such an impact in its past, Venus has been left with a thick crust but no large metallic core. Without the Moon, Earth would probably have ended up like Venus, and we would not be here.

Mars also lacks our kind of tectonic activity and a significant magnetic field, but for different reasons. It is a small planet, no bigger than the core of the Earth, so it cooled quickly. With only a small core of its own and little internal heat, it could not contain the swirling currents of molten metal needed to make a magnetic field, nor the kind of convection currents that drive continental drift on Earth. Instead, what heat there was has escaped from the interior of Mars at hot spots which have built up huge volcanoes over very long periods of time. These include the largest volcano in the Solar System, the Nix Olympica, which covers an area the size of the state of Arizona and rises 26 km above the Martian equivalent of sea level, the 'average' surface. This and other Martian volcanoes, almost as impressive, are still growing, very slowly, thanks to the residual heat below the surface.

Unlike Venus, Mars has a very thin atmosphere. It may have started out with a much thicker atmosphere – there is even evidence that it once had oceans – but a combination of its weak gravitational pull (because of its small size) and the lack of a magnetic field allowed most of the gas to escape, plunging the planet into deep freeze. The lack of a magnetic field means that particles of the solar wind can penetrate into the atmosphere, scouring it away as the eons pass, while the weak gravitational pull makes it easier for the scouring to remove the atmosphere. Venus is also being scoured in the same way, but it has plenty of atmosphere to lose and a stronger gravitational grip on it.

Without such a thick atmosphere as Venus, if it did not have a strong magnetic field the Earth would be significantly affected by the scouring effect of the solar wind; but this is not really relevant, because the Earth would only lack a magnetic field if the Moon did not exist and the planet was more like Venus. It's time to look at the other benefits of having a large moon.

A PLANETARY STABILIZER

In terms of its diameter, the Moon is more than a quarter of the size of the Earth. It has only about one eighteenth of the mass of the Earth, but this is still far larger in proportion to the mass of the planet than that of any of the other moons of the major planets of the Solar System. As a result, the gravitational influence of the Moon on the Earth is, and has been, a major influence on the development of our planet. Together with the importance of the Moon's origin for plate tectonics, the three main influences of our companion can be summed up as the three 'T's – tilt, tides and tectonics. And even the third of these owes something to lunar gravity.

Tilt refers to the amount by which the Earth leans over in its orbit. Instead of being upright, with a line through the Earth from the North Pole to the South Pole making a right angle with the plane of the Earth's orbit around the Sun, our planet is tilted at an angle of about 23.5 degrees out of the vertical. This tilt is responsible for the cycle of the seasons. The Earth always leans in the same direction in space, so as it goes round the Sun first the Sun is on the side where one hemisphere is tilted towards the Sun and it is summer in that hemisphere and winter in the opposite hemisphere, then six months later the situation is reversed. The tilt also plays a part in the rhythms of Ice Ages – more of this later.

Although the tilt of the Earth changes slightly on timescales of tens of thousands of years, it cannot vary very much, because the gravitational influence of the Moon acts as a stabilizer. If we did not have such a large moon, or if the Moon were much farther out from the Earth, the combined influence of the Sun and Jupiter (and to a lesser extent the other planets) would tug on the Earth and make it tumble in space, so that it might suddenly switch from being nearly upright to lying completely flat in its orbit ('suddenly', on this timescale, meaning in as little as 100,000 years). This kind of behaviour is chaotic, in the mathematical sense of the term, which means that small changes in the various forces acting on the Earth would produce large and unpredictable effects. Just such chaotic tumbling has happened on Mars, where there is no large moon and where the tilt can change suddenly by at least 45 degrees, and more slowly by as much as 60 degrees. But on Earth, the tilt has been essentially constant for at least hundreds of millions of years, and probably a lot longer.

It doesn't take much imagination to appreciate the effect on an incipient technological civilization if the Earth suddenly rolled over on its side, with the North Pole, say, pointing directly at the Sun. The oceans and land around the equator would freeze over, and at high latitudes each hemisphere in turn would experience a sequence of searing summers followed by freezing winters. The equatorial regions would never thaw, even when the Earth was 'side on' to the Sun in its orbit, because the shiny surface of ice and snow would reflect away most of the incoming solar heat. The tropics are, of course, home to the vast majority of species on Earth, most of which would go extinct. It seems that chaotic changes in tilt are normal for terrestrial planets, and this alone could be relevant for the emergence of a technological civilization, or any kind of complex life based on land, on a planet that 'just happens' to have a large moon.

As the Moon is slowly retreating from the Earth, this stabilizing influence will decline as time passes, which sets a limit on the window of opportunity in which a civilization like ours could have emerged on Earth. When the Moon formed, it was much closer to Earth, and has been steadily retreating as the energy of its orbital motion has gone into stirring up tides. At present it is moving outwards at a rate of about 4 cm per year, and within 2 billion years it will no longer be able to stabilize the Earth's tilt. This ties in with one of the most curious coincidences in astronomy – indeed, in science – which seems to have no explanation and is utterly puzzling. Just now, the Moon is about 400 times smaller than the Sun, but the Sun is about 400 times farther away than the Moon, so that they look the same size on the sky. At the present moment of cosmic time, during an eclipse, the disc of the Moon almost exactly covers the disc of the Sun. In the past, the Moon would have looked much bigger and would have completely obscured the Sun during eclipses; in the future, the Moon will look much smaller from Earth and a ring of sunlight will be visible even during an eclipse. Nobody has been able to think of a reason why intelligent beings capable of noticing this oddity should have evolved on Earth just at the time that the coincidence was there to be noticed. It worries me, but most people seem to accept it as just one of those things.

Tides are less worrying, because they are well understood. And they must have played a significant part in the emergence of life from the sea and on to the land. Tides on Earth are primarily produced by the gravitational pull of the Sun and the Moon – in principle there are tiny effects from other planets, but too small to be noticed. Both the Sun and the Moon cause both the oceans and the solid Earth to bulge upwards underneath them and on the opposite side of the planet (you can think of the bulge on the far side as being related to a stretching of the Earth as it is tugged towards the Moon or Sun). In between, we have low tide. On their own, lunar tides today are about twice as big as solar tides. But the two tides add together or partially cancel out at different times of the month. At New Moon and Full Moon, the Moon, Earth and Sun are in a straight line and the tides add together. This brings very high tides known as spring tides (because they 'spring up'; nothing to do with the season spring). At the quarter moons, the Sun, Earth and Moon form a right angle, and the solar effect cancels out half of the lunar effect, producing much less impressive high tides, known as neap tides. There are local variations caused by the shapes of coastlines, but in essence this means each place on Earth has two high tides and two low tides each day, as the Earth rotates under the Sun and Moon.

Even today, the ocean tides seem impressive, and in some ways it is even more impressive that tides in the 'solid' Earth have an amplitude of about 20 cm. The ground beneath your feet literally goes up and down over this range twice a day, but you don't notice because you are going up and down with it. The solar influence has been constant as long as the Earth has been in its present orbit. But when the Moon was closer to the Earth, the tides it raised, both in the seas and in the solid Earth, were correspondingly larger.

Simulations of the event in which the Moon formed suggest that it coalesced out of a ring of debris no more than 25,000 km above the Earth, compared with its distance of just over 384,000 km today. That's less than a tenth of the present Earth–Moon distance. This would have raised enormous tides in the oceans, if there had been any oceans at the time, but as it was the repeated stretching and squeezing of the solid Earth, associated with solid tides more than a kilometre in height, would have generated enough heat to keep the surface molten for some time after the impact. But the enormous amount of energy released would have seen the Moon move outwards relatively quickly, and things would have settled down enough for the Earth's crust to form (or re-form) within a million years. Even so, the heat generated by lunar tides within the Earth would have remained significant, and contributed, along with the heat from radioactivity, to the establishment of tectonic activity on Earth.

The Earth was also spinning much faster just after the impact that created the Moon – as a direct result of that impact. Tidal forces have slowed the spin of the Earth as the Moon has retreated from us. Just after the Moon formed, a day on Earth was only five hours long. At that time, instead of tides 2 metres high every 12 hours there would have been tides several kilometres high every two and a half hours. But these extreme tides did not last long. The first reasonably complex forms of plant life on land emerged from the sea a little over 500 million years ago, and in a memorable numerical coincidence there were about 400 days in the year about 400 million years ago; so the emergence of complex life from the sea occurred when tidal conditions were not dramatically different from those of today. The plants, and later animals, that made the transition onto the land could do so by spreading out from the tidal zones. First they evolved the ability to survive drying out twice a day in the intervals between high tides, then some of them developed the ability to survive above the tide line altogether. This must have been a huge evolutionary advantage, giving them the ability to spread into and colonize vast areas where there were no predators. Of course, the predators soon followed! But would it all have happened so easily without the large tides associated with our large Moon?

The more we look, the more important the Moon seems for our existence. Because of this, it's worth looking again at plate tectonics, from a slightly different perspective, and bearing in mind that plate tectonics on Earth is a direct consequence of the impact that formed the Moon and of the continuing influence of the Moon in keeping the Earth's interior hot and stirring up convection currents.

PLATE TECTONICS AND LIFE

Peter Ward and Donald Brownlee have described plate tectonics as 'the central requirement for life on a planet' not least because 'it is necessary for keeping a world supplied with water.' But the first thing plate tectonics did was to supply the Earth with land, turning a potential water world into one with oceans and continents. Without continents, there would be no land-based life to look at the stars, dig up ores to build a technological civilization, and speculate about the possibility of life elsewhere in the Universe. The relevance of plate tectonics to keeping our world supplied with water is that it plays a vital role in maintaining the temperature of the Earth in the range where liquid water can exist, and in cycling the water from the oceans onto the land and back again.

The temperature at the surface of a planet depends on how much heat it gets from its parent star, how much is reflected back into space, and how much is trapped by the planetary atmosphere (the greenhouse effect). We have a convenient indication of what the temperature of the Earth would be like if it had no atmosphere and only the first two factors came into play – the Moon is made of the same material as the surface of the Earth, it is at essentially the same distance from the Sun as we are, and it has no atmosphere. The average temperature at the surface of the airless Moon is minus 18 °C, but the average temperature at the surface of the Earth is plus 15 °C. The greenhouse effect of the Earth's atmosphere is responsible for that 33 °C difference. The size of the greenhouse effect depends on the concentration of gases such as carbon dioxide, methane and water vapour in the Earth's atmosphere (nitrogen, the main component of the atmosphere, does not trap heat in this way and does not contribute to what is sometimes referred to as the 'global thermostat'). And the concentration of these greenhouse gases is regulated largely by plate tectonics – or was, until human activities started to affect the natural cycles.

Greenhouse gases, in particular carbon dioxide and water vapour, are released by volcanic activity on any rocky planet big enough to have a hot interior. The gases come from chemical compounds in the rocky material from which the planets formed. On a planet like Venus, where there is no plate tectonics and no liquid water, these gases have nowhere to go, and build up in the atmosphere so that the greenhouse effect gets stronger and stronger as time passes. But on a planet like the Earth, where there is plate tectonic activity and liquid water, the situation is more complicated. Carbon dioxide dissolves in water, and then reacts with minerals in the rocks, in particular silicates, to produce calcium carbonate – limestone. The chemistry proceeds particularly effectively in shallow seas, less than about 4 metres deep. In effect, carbon dioxide has been turned into rock, so it can no longer contribute to the greenhouse effect; it is no coincidence that the amount of carbon dioxide in the atmosphere of Venus, which is nearly the Earth's twin in size, is roughly the same as the amount of carbon dioxide locked up in all the carbonate rocks on Earth.

But there's more. Weathering proceeds faster if the water involved is warmer. So if the Earth warms a little, more carbon dioxide is taken out of the atmosphere, reducing the greenhouse effect so that the planet cools. As it cools, weathering becomes less efficient and more carbon dioxide stays in the air, warming the world again. This is an example of what is known as a negative feedback; whichever way the temperature fluctuates, the natural tendency is to push it back towards the long-term average. Plate tectonics comes into the story because material produced by weathering is carried down to the sea and deposited as sediments on the ocean floor. There, it is carried by the oceanic conveyor belt to subduction zones where it is forced beneath the crust of a continent and melts back into the hot material below. Some of the carbon dioxide is released by this process, and gets back into the atmosphere through the volcanoes associated with subduction zones. But the mountain ranges built up by this activity encourage rainfall and weathering, making carbonate rocks on land, and carrying carbonate back into the sea. The effect of plate tectonics is to speed up the whole feedback cycle, so that it can respond quickly, by geological standards, to changes in temperature, and prevent wild swings from one extreme to the other. The process is so efficient that if the Earth were gradually moved out to the orbit of Mars, the concentration of carbon dioxide in the atmosphere would build up to about twelve thousand times its present value, with an enhanced greenhouse effect keeping the planet warm enough for liquid water and life.

All of this, though, only works effectively if the planet has a shallow layer of water, which makes it possible for land to poke above the surface. If the whole planet were covered by deep ocean, which seems entirely possible given the ease with which water was brought down to the Earth by impacts when the Solar System was young, quite apart from the fact that there would have been no land on which life forms like us could have evolved, there would have been no shallow seas in which chemical processes involving the runoff of material from the land could have laid down copious quantities of carbonates. As the temperature of the Sun increased, the world would have warmed until the water boiled away. But if there had been significantly less water and correspondingly more land area on Earth than there really is, there would have been a tendency for extreme fluctuations in temperature (because land both heats up and cools down quicker than the sea), while increased continental weathering would have drawn carbon dioxide out of the atmosphere, cooling the planet and causing extreme glaciations.

On the real Earth, life comes in to the story as well, both because it encourages weathering on land and because tiny marine creatures build their shells out of carbonates which are deposited on the sea bed when they die; the famous white cliffs of Dover, like all chalk cliffs, are made from the remains of countless numbers of tiny shells laid down in this way. The way all these processes interact on a living planet like the Earth has been investigated most thoroughly by James Lovelock, and the details can be found in his books; Wallace Broecker's *How to Build a Habitable Planet* also covers some of this ground. What matters here, though, is that although other factors are involved, without plate tectonics the Earth would not have been able to sustain a surface temperature in the range required for the existence of liquid water (which means, in the range required for the existence of our kind of life) for the billions of years it has taken for creatures like us to have evolved. It would probably have become a hot desert like Venus; just possibly it might have become a frozen world as cold as the Moon.

All this is an example of plate tectonics ensuring stability on Earth. But as far as life is concerned, its other key role is to encourage change! Imagine a planet the size of the Earth, with continents and seas but no continental drift, no mountain building, no changing climate. It would be fixed in a state of stability. Life might exist on such a planet, but each species would be exquisitely suited to its own ecological niche, with no need to change or to evolve except to become even better suited to its own way of life. There have been times in the long history of life on Earth that have, indeed, been more or less like that. But there have also been times when continents have collided, throwing up new mountain ranges, changing the pattern of rainfall, and bringing species that used to live on separate continents – in effect, in separate worlds – into contact and competition with one another. And there have been times when continents have split apart, creating different environments on each continent, separating the descendants of what used to be a single species into two populations forced to go their separate evolutionary ways as they adapt to their new homes.

Charles Darwin got his great insight into the mechanism of evolution, natural selection, from his observations of the way closely related species of birds had adapted to different conditions, forcing them to lead different kinds of life, on the neighbouring islands of the Galapagos. But if all those islands had been identical, the birds would all have been adapted to the same kind of lifestyle, and there would have been no differences between them for Darwin to notice.

It is no coincidence that the deep ocean is the part of the globe that is least affected by the changes associated with plate tectonics, and that the deep ocean is the part of the world with the least diversity of species. Nor is it any coincidence, as I shall elaborate in Chapter 8, that our own species evolved in a part of Africa that is being torn apart by tectonic activity, at a time when continental drift was changing the climate of the globe.

All in all, the case for plate tectonics being the single most important factor in making the Earth special seems secure. And as David Stevenson, of Caltech, has commented, where planets are concerned 'plate tectonics is neither mandatory nor common.' But the single most important factor in ensuring that the Earth has plate tectonics is the Moon – the impact that gave birth to the Moon meant that the Earth was left with a thin crust, and tidal heating of the interior of the Earth helped to kick start tectonic activity when the Earth was young.

Since the Moon has also acted as a planetary stabilizer, preventing the Earth from toppling over in space, and may also have shielded us from some of the pieces of cosmic debris that might have struck our planet if the Moon were not in the way, what this really means is that the single most important factor in making the Earth a suitable planet on which a technological civilization could evolve is the Moon. And moons like ours really are rare. The kind of collision in which our Moon formed, between 30 million and 50 million years after the Sun was born, must spread large amounts of dust around a planetary system. Yet when astronomers using the Spitzer infrared space telescope looked at stars roughly the same age that our Sun was when the Moon formed, they found signs of such telltale dust in only one out of four hundred candidates. Since a planet-smashing impact need not necessarily produce a large moon, the odds against are even longer than this suggests. Given the extraordinarily small likelihood of a double planet like the Earth–Moon system forming in the way it did, this reduces dramatically the already very small chance of finding a civilization like ours elsewhere in the Galaxy. But we have by no means exhausted the long list of special factors that are prerequisites for our existence.

6

What's So Special about the Cambrian Explosion?

I. Contingency and Convergence

The story of life on Earth is also the story of death on Earth. Not just individuals, but species and genera die and are replaced by others. This is a key feature of evolution by natural selection. Individuals that are well adapted to their environment thrive and leave plenty of offspring; individuals that are less well adapted die young and leave fewer offspring. If that were the whole story, a planet like the Earth would end up with a limited number of species each perfectly adapted to its own ecological niche, and in a sense evolution would stop. But it is not the whole story. The environment changes, so that old niches disappear and new ones open up. Sometimes, a species is wiped out through no 'fault' of its own, as when a large meteorite strikes the Earth. But what are catastrophes from the point of view of some species are opportunities for others, which diversify and radiate to fill the gaps left by the species that have gone extinct. Just how this happens, and what the long-term consequences are, has been the subject of a fierce debate among the experts. Stephen Jay Gould, of Harvard University, argued most vociferously from one extreme, while Simon Conway Morris, of Cambridge University, has responded from the other. To an outsider, the truth (as is so often the case) seems to lie somewhere between the extremes, in a merger of key points from each of their scenarios.

THE CAMBRIAN EXPLOSION

The background to this debate is the sudden (by geological standards) proliferation of complex organisms, modern multicellular animals, in the fossil record about 570 million years ago. This explosion of multicellular forms is so significant that it is used as a marker in the geological record – the most significant 'date stamp' in all of geology. It is the beginning of the geological period known as the Cambrian, which lasted until about 485 million years ago. The Cambrian period is the start of the Palaeozoic era, which lasted until about 225 million years ago and has been succeeded by the Mesozoic era (225 to 65 million years ago) and the Cainozoic era (from 65 million years ago up to date). The boundaries between eras are defined by significant changes in the flora and fauna that inhabited the Earth. But everything before the Cambrian explosion, all the first 3.5 billion years of geological time, is regarded as a single era, the Precambrian, in which no comparable dramatic changes occurred and the oceans swarmed with single-celled life.

In terms of the prospect of finding intelligent life elsewhere in the Galaxy, the Cambrian explosion is disappointingly late – at least 3 billion years after the emergence of life on Earth. Leaving aside the more circumstantial evidence for life 3.8 billion years ago, the fossil remains of actual cells have been recovered from rocks dated at around 3.6 billion years old. Since the essentially identical descendants of those single-celled organisms still thrive on Earth today, they are arguably the most successful species in the history of life on Earth. For at least 2.2 billion years (2.4 billion if we accept the circumstantial evidence), all life on Earth consisted of these simple cells, known as prokaryotes, which are essentially little bags of jelly containing the chemicals of life (such as DNA and proteins) but without the nucleus and other internal structures that characterize the so-called eukaryotic cells of creatures such as ourselves.

Eukaryotic cells are much bigger than prokaryotic cells, and as well as the nucleus which houses their DNA they contain other structures, such as the chloroplasts that carry out photosynthesis in plants and the mitochondria that use chemical reactions to generate the energy needed by the cells in your body. Lynn Margulis, of the University of Massachusetts Amherst, established that this structure is a result of formerly free-living prokaryotes getting together to live in symbiosis inside a single cell wall. This must have conferred an evolutionary advantage, or we would not be here to discuss it. But eukaryotic cells only appear in the fossil record some 1.4 billion years ago. Even then, our kind of organisms did not emerge at once. The time from the appearance of eukaryotes to the explosion of multicellular forms at the start of the Cambrian is nearly one and a half times as long as the time from the Cambrian explosion to the present day.

There is some evidence of two kinds of multicellular life forms in strata from the late Precambrian or very early Cambrian, the 100 million years or so immediately before the Cambrian explosion. But these do not look like the ancestors of modern animals. One kind, known as Ediacara, seem to have been flat, soft-bodied creatures, divided into sections that have been described as resembling the structure of a quilt, or an air mattress. Gould argues that such organisms would be unable to evolve complex interiors, and that if the Ediacara had continued to dominate, animal life on Earth would have remained 'permanently in the rut of sheets and pancakes – a most unpropitious shape for self-conscious complexity as we know it'.

The first evidence of creatures with shells comes from the earliest Cambrian; but they too do not seem to be the ancestors of modern animals – nor, indeed, to be part of the same branch of life as the Ediacara. They are called the Tommotian, and are summed up by palaeontologists as 'small, shelly fauna'. Both the Ediacara and the Tommotian disappeared when the Cambrian explosion introduced a huge variety of new body types into the fossil record, including the basic phyla, the main divisions of the animal kingdom as we know it. So the evolutionary road that leads to us starts with the Cambrian explosion. But what route did that road take?

THE BURGESS SHALE

A great deal of what we know about animal life in the early Cambrian comes from studies of fossils found in the Burgess Shale, a geological formation in the Canadian Rockies. These remains were laid down about 530 million years ago in the sediments at the bottom of a cliff in a shallow tropical sea – animal life would not begin to move on to land for another hundred million years, and vertebrates only moved on to land about 380 million years ago. The Burgess Shale is particularly important because it has, unusually, preserved the remains of soft-bodied creatures, not just the bony or shelly bits that fossilize easily. But the more familiar kinds of fossils are also found there, in the same proportions that they occur elsewhere in strata of the same age where traces of the soft animals are not preserved. These 'hard' fossils make up only 5 per cent of the fauna in the Burgess Shale, which indicates just how restricted palaeontologists would be if they only had these bony and shelly bits on which to base their studies. It seems that the Shale provides a complete snapshot, the best of its kind, of animal life at the time. As Conway Morris stresses, the bony bits, although in a minority, are just the same as the bony bits found in strata of the same age in other parts of the world, telling us that the members of the Burgess Shale fauna are not the product of some freak evolutionary development in a region cut of from the rest of the globe, but 'are part of the mainstream of Cambrian life'. This fauna has become widely known through Stephen Jay Gould's book *Wonderful Life*, which is based on the reconstructions of the animals made by Simon Conway Morris and his colleagues in Cambridge, although Gould and Conway Morris disagree almost completely on how to interpret this evidence in evolutionary terms.

The evidence itself is clear enough. The creatures alive in the early Cambrian came in an enormous variety of shapes with bizarre anatomical features, many bearing little resemblance to remains found earlier or later in the fossil record (earlier or later than the Cambrian, that is). More than 70,000 specimens have been found, but to take just a few examples: one, dubbed Nectocaris, looks like a vertebrate with a shell (or if you think about it the other way, like a crustacean with fins); another, so weird that Conway Morris named it Hallucigenia, had a long body supported on seven pairs of stilt-like legs, with a lump at one end that might be a head, seven long tentacles waving above the long body, six short tentacles at the presumed tail end of the beast, and the body bending upwards into a tube beyond this cluster of tentacles; and Opabinia, a creature with five eyes and a kind of trunk ending in a claw, is so bizarre that when the palaeontologist Harry Whittington first presented a drawing of it to a scientific meeting in Oxford the audience burst out laughing.

If anything, though, the speed with which the Cambrian explosion occurred is even more important, in the context of understanding the reasons why we are here, than the variety of animals it produced. Even in the nineteenth century, it was clear to geologists that something odd had happened at the beginning of the Cambrian – which is why, of course, they lumped all earlier Earth history together under the label Precambrian. At first, they could find no traces of life from the Precambrian, something which was a cause of profound concern to Charles Darwin, among other people. He discussed the puzzle in the *Origin*, and admitted that the apparent appearance of fully formed complex life forms at the start of the Cambrian was the most compelling argument that could be used against his theory. But we are now able to trace the origin of life back to the first prokaryotes, removing that argument. We are left with the puzzle of how and why complex forms evolved very quickly from simpler ancestors, but not literally overnight, or in the course of a week. To a palaeontologist, the term 'explosion' conjures up an image of a diversification of life that took place over an interval of a few million years. Conway Morris describes how it is possible, in the Flinders Ranges of Australia, to walk along a gorge where strata have been tilted by geological forces so that instead of lying on top of each other they lie side by side. A walk along the gorge is like a walk though time, passing, first, layer upon layer of rock containing Ediacaran fossils. Then, the Ediacarans disappear and are 'replaced by something much more significant: skeletons. Here is the most obvious manifestation of the Cambrian explosion.'

This is a dramatic indication of the reality of the Cambrian explosion. And one of the important features of the Burgess Shale is that because it contains bony bits as well as the fossilized remains of soft-bodied creatures, it gives us confidence that the diversification we see in strata where only the bony bits are found is representative of what was going on in the overall fauna. The Cambrian explosion was a genuine evolutionary event, the most dramatic in the fossil record, a time for which there is, in Conway Morris's words, 'compelling evidence for a profound change in anatomical, ecological, and neurological complexity'. I shall look at what might have triggered this unique event in the next chapter; but first, how did this diversity of Cambrian life evolve into the modern living world, a world in which technological civilization has arisen?

CONTINGENCY

There's a popular image of the way evolution has produced the diversity of life on Earth which starts from a single cell and branches repeatedly, like a tree which grows upward from a single stem and spreads out into a glorious canopy above. In view of the billions of years for which the only forms of life on Earth were single-celled creatures floating in the sea or forming mats on the sea bed, the 'stem' would have to be very long indeed in proportion to the height of the tree; but you get the picture. More simplistically, this becomes the image of a cone standing on it point, with the broad end uppermost and the variety of life radiating out from the original ancestor. The variety is produced by species spreading out into new ecological niches and adapting to different environments, and the diversity increases as time passes.

But this view of the evolution of diversity has been challenged, most notably by Stephen Jay Gould, who promoted his alternative view in his influential book *Wonderful Life*. Unfortunately, 'influential' does not necessarily mean 'correct'; but Gould's argument is so well known that it is important to address it and explain why it is wrong.

Gould literally turns the image of diversity spreading like an inverted cone getting wider as it stretches away from its pointed base on its head. He says that a better analogy is with a cone standing on its base and getting narrower as it shrinks down towards the tip. On this picture, the base of the cone corresponds to the explosion of diversity that occurred at the beginning of the Cambrian and is recorded in the Burgess Shale, but since then diversity has been decreasing, not increasing, and many of the early experiments with body plans and perhaps even lifestyles have disappeared. This does not mean that the number of species has been decreasing. Palaeontologists agree that the number of species has increased as time has passed. But if there is a large number of related species, they can all be regarded as variations on a single theme, defined by a single basic body plan – all mammals, for example, have the same basic structure. It is the number of basic body plans that has decreased, according to Gould.

Adapting the 'tree of life' analogy, his version is more like that of a Christmas tree, with a trunk corresponding to the monocellular monotony of life in the Precambrian, a sudden spreading out of diversity at the beginning of the Cambrian, and then a tapering towards a point. Whatever the reasons for the appearance of multicellular animals at the start of the Cambrian, all kinds of complex life forms appeared in a flourish of evolutionary experimentation which has since been reduced to a trace of its former glory – as Gould puts it, 'the maximum range of anatomical possibilities arises with the first rush of diversification.' Most of these early experiments then disappear as life settles down, producing many variations but on only a few remaining themes. This is best seen, Gould says, in terms of the number of phyla, the main division of species within the animal kingdom. At present, there are about thirty-five phyla (the biologists cannot agree on exactly how many, because the boundaries are somewhat arbitrary). But Gould claims that many of the animals found in the Burgess Shale do not belong in any of the modern phyla. In a key passage of his book, he says, 'the fifteen to twenty unique Burgess designs are phyla by virtue of anatomical uniqueness. This remarkable fact must be acknowledged with all its implications . . . ' and, according to Gould, those fifteen to twenty phyla have no descendants alive today

– implying that perhaps twenty out of an original fifty-five phyla, almost half, have disappeared.

This is a contentious claim; but let it ride for the moment. Gould goes further. He says that the disappearance of so many phyla has not been simply a result of what is usually called 'Darwinian' evolution, with the best-adapted species (the 'fittest', in the sense of best fitting in to their environment) surviving and the rest becoming extinct. Instead, he says that there is an element of chance – or as he puts it, 'contingency' – in determining which species and which phyla survive and which ones go extinct. In other words, a whole phylum might disappear from the face of the Earth as much as a result of bad luck as as a result of bad genes. The kind of bad luck that would cause such extinctions might be an impact from space, or a change in climate caused by continental drift. As Gould puts it, 'even if fishes hone their adaptations to peaks of aquatic perfection, they will all die if the ponds dry up.'

It is the idea of contingency that leads to the most powerful image in Gould's book – the idea of replaying the 'tape' of life on Earth starting out once again from the fauna that existed at the time the Burgess Shale was being laid down. What if we could rewind evolution to start from that time again? Would we end up with a world very similar to the Earth today, with creatures like ourselves living in a world with animals and plants like our own? Or would our putative ancestors be among the phyla that got wiped out, by accident rather than by design, leaving a world populated by, among others, strange species descended from the phyla that Gould believes went extinct in our version of history? His answer is unequivocal: 'any replay of the tape would lead evolution down pathways radically different from the road actually taken ... Alter any early event, ever so slightly and without apparent importance at the time, and evolution cascades into a radically different channel.'

This leads Gould to the conclusion that the evolution of human beings, starting out from the Cambrian explosion, was an overwhelmingly improbable event. And he backs this up by pointing to examples in the real world today which suggest that the rise of the apes (or, at least, one kind of ape) to intelligence was a far from foregone conclusion.

There's a common view of the rise of the mammals, in the context of the history of life on Earth, which says that the dinosaurs dominated the planet for hundreds of millions of years, while mammals existed as small creatures scurrying about in the undergrowth or burrowing in the ground, perhaps nocturnal, keeping out of the way of the dinosaurs. It was only when the dinosaurs died out, some 65 million years ago, that this opened the way for mammals to spread out and diversify, occupying the ecological niches recently vacated by the dinosaurs, and leading inevitably to the 'rise' of the apes and the emergence of ourselves. But Gould points out that 50 million years ago the rise of the mammals would not have seemed inevitable to any alien biologist visiting the Earth. At that time, gigantic carnivorous birds, close relatives of the dinosaurs, roamed across Europe and North America. Some of them were as much as two metres tall, with massive heads, powerful beaks, and viciously clawed feet, but only vestigial wings. They were more like pocket versions of Tyrannosaurus Rex than like modern songbirds. Birds and mammals were rivals to succeed the dinosaurs, and we don't know why the mammals succeeded and the birds failed. Gould thought it was just by chance, what he called contingency, not through any evolutionary superiority of the mammals, and he backed this argument up with examples from South America.

Fifty million years ago, South America was an island continent, which only got joined to North America a few million years ago. In South America, birds did become the dominant carnivores, although these species largely died out before the south was invaded by mammals from the north across the Isthmus of Panama. Again, we don't know why – neither why the birds came to dominate South America, nor why they went extinct. But the fact that mammals became the dominant carnivores in one part of the world and birds became the dominant carnivores in another suggested to Gould that in either case it was as much a result of luck as of evolutionary superiority.

But there's something else about those South American carnivorous birds that gives a different insight into evolution. Although they were not closely related to the pocket T. Rex that roamed across Europe and North America, they stood nearly three metres tall, and also had big heads, stocky necks, vestigial wings, and viciously clawed feet. This is an example of convergent evolution – the same evolutionary pressures that made this kind of body plan successful in Europe and North America also made it successful in South America. To Gould, this convergence is an incidental feature, of only secondary importance. But to Conway Morris, convergence lies at the very heart of the story of why we are here.

CONVERGENCE

Much of Gould's argument rests upon the claim that many phyla present in the Burgess Shale have no living counterparts today. Now, Gould didn't actually analyse the fossils from the Burgess Shale himself. As he acknowledged, his interpretation is based on the reconstructions carried out by Simon Conway Morris and his colleagues. But what is less widely appreciated is that this interpretation is based on early work by these people. At the time Gould was writing *Wonderful Life*, in the second half of the 1980s, these people were still beavering away at their analyses and interpretations of the fossil remains. After Gould's book appeared, more information emerged from the Burgess Shale itself (including new insights into the archetypal inhabitant of the Burgess Shale, Hallucigenia), and new discoveries were made in strata of the same period from China, Greenland and other places around the world. All of this encouraged Conway Morris to write his own book, *The Crucible of Creation*, which appeared at the end of the 1990s. So what does he make of the evidence?

According to Conway Morris, 'the evidence to date does not support [Gould's] metaphor of an "inverted cone of life".' Although it seemed at first sight difficult to fit many of the bizarre-looking fossils from the Burgess Shale into known groups, such as phyla, on closer inspection the Cambridge team became convinced that rather than representing extinct phyla, the weird and wonderful fossils from the early Cambrian actually provide insight into how the present phyla evolved. In popular language that Conway Morris would probably disapprove of, they represent 'missing links' rather than separate branches of the evolutionary tree. He even finds a place in his scheme for the Ediacarans, which he interprets as members of the phylum Cnidaria, which includes modern jellyfish.

Conway Morris rejects both the old idea of diversity (in my sense of the term) increasing as time passes like a cone standing on its tip, and Gould's idea of diversity decreasing as time passes like a cone standing on its base or a Christmas tree. And he also rejects the possibility that there was an explosion of diversity in the early Cambrian but that since then there has been little change. Instead, he argues that there was indeed a proliferation of animal life at the start of the Cambrian, and that since then there have been long intervals in which there was little change in diversity, broken by short intervals in which diversity has increased, although never on the same dramatic scale as at the start of the Cambrian – 'occasional innovations in evolution, the consequences of which ripple through the biosphere and so drive times of rapid change'. These lesser explosions in diversity are often (perhaps always) linked with changes in the terrestrial environment, associated with geographical changes caused by continental drift, impacts from space, and so on; this is an important clue to the reason for the Cambrian explosion itself, which is discussed in the next chapter.

One of Conway Morris's themes is the continuity and interrelatedness of life on Earth, and what this tells us about human origins. At the most basic level of biochemistry, all the cells in your body (and every other living cell on Earth) operate in essentially the same way as the cells of 'primitive' bacteria, compelling evidence that all life on Earth today has evolved from a single ancestor. The most significant human characteristic, our brains, although large and impressive, are constructed in essentially the same way as the brains of primitive fish, showing that this structure evolved at least 500 million years ago. And the structure of what became the pentadactyl limb, with five fingers or toes on each appendage, can be seen in the fossil remains of fish found in rocks 370 million years old from Greenland.

As these examples show, it is astonishing how much variety evolution can produce from a single type of body plan. Taking an example a little farther from humans, the snails in your garden, limpets, oysters and even the free-swimming (or bottom-walking) octopi are all members of the mollusc phylum, and have all evolved from a single ancestral mollusc. The octopus is a particularly interesting example, because it has evolved an advanced form of brain and an effective 'camera eye' quite independently of the evolutionary path that has led to our brains and our eyes. As Conway Morris says, in evolutionary terms, 'There are not an unlimited number of ways of doing something. For all its exuberance, the forms of life are restricted and channelled.' This is where convergence comes in, suggesting that much more than the luck of the draw was involved in producing our kind of intelligence.

We've already met one example of convergence, the way the killer birds of South America and those of Europe and North America evolved similar bodies in order to achieve success in similar ecological niches, at about the same moment of geological time. There are other examples which are widely separated in time. The marine species known as ichthyosaurs swam in the oceans of the world between about 245 million years ago and 90 million years ago, thriving in particular in the Jurassic period, between about 200 million years ago and 145 million years ago. They were air-breathers that had evolved from land animals that returned to the water, they grew to a length of two or three metres, and they even gave birth to living young, instead of laying eggs. They looked, and presumably behaved, like modern dolphins. The same evolutionary pressures that made an ichthyosaur body and lifestyle successful in the Jurassic made the dolphin body and lifestyle successful in the present geological period, the Quaternary.

Conway Morris highlights some other examples. A sabre-toothed big cat, similar to the well-known sabre-toothed tiger, evolved in South America from marsupial stock – in spite of its outward appearance, it was more closely related to the kangaroo than to living tigers. And Australia, home of the kangaroo, is also home to a marsupial mole, which, like its mammalian counterpart, is 'also equipped with powerful forelimbs for digging and with only weak eyes in keeping with its subterranean existence'. Using these and other examples (which are elaborated in his book *Life's Solution*) he emphasizes that 'again and again we have evidence of biological form stumbling on the same solution to a problem.' If the tape of life really could be replayed from the Cambrian explosion, although 'the evolution of the whale ... is no more likely than hundreds of other end points, the evolution of some sort of fast, ocean-going animal that sieves sea water for food is probably very likely and perhaps almost inevitable.' John Maynard Smith, who was an aeronautical engineer before he turned to evolutionary biology, once commented to me in a similar vein that what he called 'engineering design' restricts the possible variety of biological form and makes it no surprise that, for example, the camera eye has evolved more than once (in fact, at least six times). The number of evolutionary possibilities is limited, and possible solutions to evolutionary problems have usually been arrived at several times. This, of course, is powerful confirmation that life does evolve by adapting itself to the conditions in which it finds itself.

But is the evolution of an intelligent species that mines metallic ores and builds radio telescopes and spaceships either inevitable or likely? Pushing the convergence argument to its limit, Conway Morris suggests that, starting out from the Cambrian explosion, intelligence was almost inevitable. This is the exact opposite of Gould's view that intelligence was extremely unlikely. But Conway Morris's argument is not much comfort to anyone hoping to make contact with ET, because he believes that life itself 'was the product of a freak and extraordinarily rare event' that took place early in the history of our planet. In his view, once you have life, intelligence is inevitable; but Earth may be the only planet where there is life. In my view, both contingency and convergence come into the story of life on Earth and the emergence of human beings. But that doesn't provide the ET enthusiasts with any more grounds for optimism.

THE THIRD WAY

In the real world, both contingency and convergence have played their parts in producing a technological civilization. The classic example of contingency is the 'death of the dinosaurs', some 65 million years ago, now widely recognized as resulting from the impact of a large meteorite, a rock some 10–12 km across, with the Earth. More of this in the next chapter; the most salient point here is that for well over a hundred million years before that impact dinosaurs had dominated the large-animal niches on Earth, while mammals existed as insignificant members of the global ecology. It was the death of the dinosaurs that gave mammals their chance, even if that chance may also have involved opportunities for birds. As for convergence, there is no better example than the ichthyosaur and the dolphin.

But the extinction of many species of life on Earth that we call the death of the dinosaurs, and which is more properly called the 'terminal Cretaceous event', since it marked the end of the Cretaceous period of geological time, was not a unique event. The fossil record shows that there have been many occasions since the Cambrian when, for whatever reason, large numbers of species went extinct. The most extreme of these mass extinction events are known as the 'Big Five', and the terminal Cretaceous event was the most recent of these. But it was not the biggest.

The Big Five are defined as mass extinction events in which at least 65 per cent of marine animal species were wiped out – marine animals are chosen as the benchmark because their fossils tend to be preserved more effectively than those of land animals. But in a sense it is easier for a phylum to survive than a species – before a mass extinction, a particular phylum may have many member species; after the event, no matter how many species are wiped out if one member of the phylum has survived the phylum still exists. The most extreme of the Big Five events marked the end of the Permian period, about 225 million years ago, when 95 per cent of the marine animal species disappeared. The evolutionary slate was almost literally wiped clean of complex, multicellular organisms – and it was in the aftermath of this event that the dinosaurs rose to prominence. The other members of the Big Five marked the end of the Ordovician period, 440 million years ago, the end of the Devonian, 365 million years ago (just before vertebrates, our direct ancestors, moved on to land), and the end of the Triassic, 210 million years ago.

There is no obvious pattern to the extinctions, although the terminal

WHAT'S SO SPECIAL ABOUT THE CAMBRIAN EXPLOSION? (1)

Cretaceous event can be definitely linked to an impact from space. But what has happened before can surely happen again, and there is no reason to think that *Homo sapiens*, or any young civilization emerging on another planet, will be immune from such disasters, any more than the dinosaurs were. Yet another factor limiting the chances of a technological civilization developing to the point of interstellar travel, or even interstellar communication.

The best guess is that many of the Big Five extinction events, and possibly some of the lesser extinctions seen in the fossil record, resulted from changes in climate associated with the changing geography of the globe caused by continental drift. The terminal Permian event, for example, occurred when all the land of the planet was locked up in one supercontinent, Pangaea, which extended from pole to pole. This reduced the availability of shallow seas where marine organisms could thrive, and led to the continental interior, far from the ocean, drying out.

Palaeontologist Steven Stanley, who has made a particular study of extinctions, says that global cooling has probably been the main reason for such events, because of 'the relative ease with which a change in global temperatures can eliminate myriads of species'. In an Ice Age, high latitudes become uninhabitable, and such habitats as remain shrink towards the equator. If the tropics today ceased to have 'tropical' climate as we know it, huge numbers of species would die. As palaeo-anthropologist Richard Leakey has put it, evolution 'cannot anticipate future events'. But he has also described mass extinctions as 'a major creative force in shaping life's flow ... mass extinctions do not merely reset the clock of evolution, jolting it back for a while; they change the face of the clock. They create the pattern of life.'

The most dramatic change in the pattern of life occurred at the beginning of the Cambrian. It is now clear that there was widespread life in the Precambrian seas, and that it had existed for 85 per cent of the time from the birth of the Earth to the present day. If the environment had not changed, it is highly unlikely that the fauna would have changed dramatically. But extraordinary events require extraordinary explanations. The Cambrian explosion is the most extraordinary event in the fossil record, and it is clear that something extraordinary must have triggered it. With Stanley's words in mind, it is natural to look for an extraordinary climatic event as that trigger.

Writing in 1989, Gould summed up the central mysteries of life's history as: (1) Why did multicellular life appear so late?; and (2) Why do these anatomically complex creatures have no direct, simpler precursors in the fossil record of Precambrian times? Ten years later, Conway Morris, with a slightly (but only slightly) different perspective, says that 'we can be fairly sure that any pre-Ediacaran animals would have been tiny, only a few millimetres long ... What later triggered their initial emergence as the Ediacaran faunas, and subsequently the even more spectacular Cambrian explosion, remains a significant topic for debate.' A further decade down the line, it seems highly likely, and is now widely recognized, that a series of dramatic environmental changes on our planet around 600 million years ago provide at least a partial resolution of these puzzles; I believe that these events can themselves be linked to changes occurring across the inner Solar System at that time. One of the key reasons why we are here may be written on the face of Venus.

7
What's So Special about the Cambrian Explosion?
II. Hothouse Venus/Snowball Earth

After the initial appearance of life itself, the two most significant events involving life on Earth were the emergence of eukaryotic cells, about 2.5 billion years ago (more than a billion years after the emergence of prokaryotic cells) and the explosion of multicellular life forms, leading to an abundance of animal life, about 600 million years ago. There is compelling evidence that these events occurred in the aftermaths of the two most severe environmental changes that the Earth has experienced since the formation of the Moon, Ice Ages so extreme that even the tropics froze. One such juxtaposition might be dismissed as a coincidence; but finding the two biggest evolutionary developments each occurring after one of the two biggest environmental catastrophes clearly tells us that we are seeing cause and effect at work. As Lady Bracknell might have put it: to experience one evolutionary leap after an ice event may be regarded as a coincidence; to experience two looks like a pattern. And although we cannot know the exact mechanism by which an environmental disaster triggers a burst of evolutionary activity, this clearly matches the pattern we see on a lesser scale associated with lesser extinctions over the past few hundred million years.

AFTER THE DEEP FREEZE

These 'Snowball Earth' events are recorded in the marks left by glaciers on ancient rocks, showing that on both occasions the entire land surface of the planet was frozen, with other evidence suggesting that most, if not all, of the ocean was also covered in ice. This is a dangerous state for our planet, and another hazard for life. Once the Earth is frozen, it reflects away so much of the incoming heat from the Sun that it is extremely difficult for the ice to thaw. If ice covers the land, quite apart from the cold there is only limited space available for life (not that there was life on land 2.5 billion years ago); and if ice covers the sea, sunlight cannot penetrate into the water to permit photosynthesis or drive other energy-dependent processes. Life struggles on in isolated puddles or slushy lakes here and there, producing diversity as evolution drives cells to adapt to the environmental conditions in their own little puddle. Quite possibly the eukaryotic cell was 'invented' only once, in one slushy pool, and spread out from there when the Earth thawed.

But why would such a Snowball Earth thaw? The only possibility seems to be a buildup of greenhouse gases, in particular carbon dioxide, which would trap heat from the Sun and eventually raise the surface temperature enough for the ice to begin to melt. Once that threshold was reached, as the ice melted back the amount of heat reflected into space would diminish, encouraging further warming in a runaway process bringing an end to the snowball state.

This is very much in line with what we understand about the way the Earth's temperature is regulated today. Greenhouse gases such as carbon dioxide are released by volcanoes and enter the Earth's atmosphere. But carbon dioxide dissolves in water, and as the water trickles through and over rocks chemical reactions associated with weathering incorporate some of the carbon dioxide into rocks such as limestone – including the stalactites and stalagmites found in limestone caves. Some living creatures also take up carbon dioxide from the air and incorporate it in their shells. If the planet warms a little, weathering is increased and more carbon dioxide is taken out of the air, reducing the greenhouse effect and making the Earth cool. If the planet cools, weathering is less effective (not least because there is less rain) and carbon dioxide builds up in the air, so the Earth warms. At present (at least, before the intervention of human activities) these negative feedbacks have acted to stabilize the temperature of our planet within

WHAT'S SO SPECIAL ABOUT THE CAMBRIAN EXPLOSION? (II)

a fairly narrow range. But during a Snowball Earth event, although volcanoes would continue to pump carbon dioxide out into the atmosphere, there would be very little weathering, and the greenhouse effect would increase until the critical point at which the ice began to melt was reached. As the warming kicked in and the ice retreated, the amount of incoming solar heat reflected back into space would decline, encouraging further warming and providing more homes for life, which would explode out from its restricted sites to once again take over the world (or, at least, the oceans). The whole meltback process would take a few million years – perhaps as long as 20 million years.

With life extending out across the ocean during this interval and with both a plentiful supply of carbon dioxide in the air and nutrients being carried down off the land by the returning rains, this period of meltback would see a burst of wild photosynthetic activity, producing a bloom of organisms across the oceans, like the green scum often seen on ponds, and releasing huge amounts of oxygen into the air as a result. The direct effect of the oxygen 'spike' associated with the first Snowball Earth event can be seen in rocks 2.5 billion years old from around the world in which huge deposits of iron oxides were laid down as the free oxygen in the air reacted with iron compounds – in effect, the world rusted as it warmed out of the snowball state. The presence of free oxygen in the air for the first time also forced evolutionary changes to occur – some organisms couldn't cope and died, others adapted, learned to use oxygen, and thrived. From our own perspective, though, the key thing about the first Snowball Earth event is that it gave rise to eukaryotes. Without that big freeze, we would not be here.

But why did the Earth freeze 2.5 billion years ago? The short answer is, we don't know. The event occurred so long ago that it is difficult to find any evidence for causes. One partial explanation is that plate tectonics – continental drift – played a part. When the Earth was young, there was very little land, and without land there was nowhere for snow to settle and build up into ice sheets. Snow falling on land starts reflecting away incoming heat as soon as it settles; snow falling on the ocean is simply going to melt, unless the sea freezes. And that requires a much greater drop in temperature than you need to make snowfall. But that only explains why the Snowball Earth didn't happen sooner. The exact trigger for the event of 2.5 billion years ago remains a mystery.

There is no doubt, though, that plate tectonics played a part in the events that followed the first Snowball Earth. One reason why life flourished in the interval between the two Snowball Earth events is that by then the amount of dry land on Earth was beginning to approach the area we see today. There was, as yet, no life on the land; but continents are edged by shallow water, and shallow water, with plenty of sunlight available for photosynthesis and plenty of nutrients being washed down from the land, is an ideal place for life to thrive – as the Burgess Shale testifies. But because of the lack of data from the time, the experts cannot yet agree on the exact arrangement of the continents at the time the second phase of global glaciations set in, although this event, referred to by geologists as the Sturtian glaciation, has now been accurately dated to just over 700 million years ago.

The Sturtian glaciation lasted for at least 5 million years, and extended to the equator. But since eukaryotic life survived this deep freeze, the Earth cannot have been literally frozen solid. There must have been patches of open water or slushy pools accessible to sunlight, and this time the proliferation that followed the thaw saw, among other things, the emergence of the first animals.

Joseph Meert, of the University of Florida, and Trond Torsvik, of the University of Oslo, have reviewed the limited data that are available, and conclude that between about 800 million and 700 million years ago most of the continental surface of the globe seems to have been concentrated at low latitudes. In itself this may have encouraged the world to cool. There was no vegetation on land at the time, and bare rock reflects away incoming solar energy more efficiently than the sea does, so land at low latitudes does make the planet cooler. But can it cool the planet enough to freeze the entire globe? This seems particularly unlikely, since the other side of this geographical coin is that with all the land near the equator, the polar regions were easily accessible to warm ocean currents from lower latitudes, making it hard to freeze the sea even in winter. Once the continents were covered in snow and ice, though, the extra reflectivity concentrated in just the right place to reflect most of the Sun's incoming heat could cool the globe sufficiently to do the job, plunging it into a Snowball Earth state. Computer models show that once sea ice develops within 30 degrees of the equator it spreads to higher latitudes and the whole of the Earth freezes. But it starts from low latitudes, paradoxical though that may seem. What's needed is a trigger, some outside influence that can tip the temperature balance far enough for snow to fall at the equator. It sounds unlikely; but something must have done the trick, and in the absence of hard evidence there have been several more or less respectable speculations. I'll mention just two – one of the oldest, and my favourite.

TIPPING THE BALANCE

The first speculation is that the Earth may have been more tilted at the time of the Snowball Earth event – that it had a high obliquity. This idea actually goes back to the 1970s, as a proposed explanation for Sturtian glaciations, and predates the discovery that the entire Earth was in the grip of ice at that time.

High obliquity, with the Earth tilted at more than about 55 degrees out of the vertical, is actually a good way to cool the tropics. Under such conditions, over the course of a year the poles receive more solar heat than the equator does, and if there were any land at the poles the summer temperature there might exceed the boiling point of water. This immediately raises the question of how ice could extend to the poles under such conditions, even with glaciers in the tropics. It might work as an explanation for isolated tropical glaciation, but not for a Snowball Earth event. But let that pass; there are plenty of other difficulties with the model to worry about.

One problem is how the Earth got to be so tilted some 700 million years ago, and how it then shifted into a more upright position. After all, with the present tilt of about 23.5 degrees, the Moon acts as a stabilizer, preventing any such extreme changes in obliquity. I've suggested that that is one of the reasons we are here. But that isn't quite the whole story. It turns out that if the impact in which the Moon formed left the Earth with a tilt of between 60 degrees and 90 degrees, then it could have wobbled about chaotically within those limits, even allowing for the presence of the Moon. So in order to preserve the obliquity explanation of the Sturtian glaciations you have to invoke a flip of the planet into a more upright position about 600 million years ago, settling it into the region where the stabilizing force of the Moon could do its work.

Just how this could have happened is a mystery. One suggestion has been that if, while the Earth was highly tilted, most of the continents drifted together into a single land mass over one of the poles, it could upset the balance of the Earth, and lever it into its present position, over an interval of tens of millions of years. But although there was indeed a concentration of land around and over the South Pole at the end of the Precambrian, this highlights what is probably the biggest flaw in the argument that high obliquity causes Snowball Earth events. Changes in obliquity are slow. But Snowball Earth events both begin and end abruptly, by geological standards. Which leads to my preferred explanation for the trigger for the big freeze that preceded the Cambrian explosion. It also has its flaws, but no more than many of the proposals put forward to explain Snowball Earth, and a lot fewer than some of them. To put it in perspective, look at the impact that brought an end to the Cretaceous period of geological time, and what this tells us about the risk to life on Earth from impacts from space.

FROM WITHOUT OR WITHIN?

Evidence that the terminal Cretaceous event was triggered by the impact of an asteroid with the Earth began to emerge in the early 1980s, when Louis and Walter Alvarez discovered a thin layer of iridium in rock strata 65 million years old at widely separated sites around the world. Iridium is very rare on the surface of the Earth, but is much more common in some kinds of meteorite. They calculated

WHAT'S SO SPECIAL ABOUT THE CAMBRIAN EXPLOSION? (II)

that the impact of such an object, about 10 km across, could have spread the right amount of iridium around the globe in the cloud of debris blasted out by the impact. The idea was greeted with scepticism at first, but more traces of debris from an impact occurring at the time of the death of the dinosaurs began to turn up as geologists searched for evidence for or against the hypothesis. To most people, the clinching evidence came in the early 1990s, when a geological feature at Chicxulub in the Yucatán Peninsula of Mexico was identified as the remains of an impact crater in exactly the right place and with exactly the right age to do the job.

There remained one doubt. Iridium does exist within the Earth, and it is brought to the surface by volcanic eruptions. It still seemed just possible that volcanic eruptions on a truly massive scale could have put the iridium layer in place, and such eruptions would, of course, be bad news for life on Earth. There was even a candidate for the job. Huge eruptions at roughly the right time spread vast amounts of lava across what is now west-central India, forming a structure known as the Deccan Traps. This is one of the largest volcanic features on Earth, more than two kilometres thick and covering an area of more than a million square kilometres. But it is not a unique feature. The Siberian Traps were produced in one of the largest known volcanic events since the start of the Cambrian, which lasted for a million years and spanned the Permian–Triassic boundary, about 250 million years ago. This 'coincided' with the terminal Permian extinction event, which killed some 90 per cent of species existing at the time; in fact the eruption of the Siberian Traps almost certainly caused this extinction.

So it is reasonable to implicate the formation of the Deccan Traps with the extinction at the end of the Cretaceous, and it is certain that such events have played a part in the history of life on Earth – another hurdle to be overcome before a technological civilization can emerge on a planet like the Earth. And it also seems possible that life at the end of the Cretaceous was already under stress because of the environmental changes associated with the formation of the Deccan Traps when the meteorite struck. All of these possibilities encouraged a great deal of work over the best part of the next two decades, culminating in a meeting of forty-one geologists, palaeontologists and other researchers who reviewed all the data and published their conclusions in the journal *Science* in 2010. They found that the volcanic activity lasted for about 1.5 million years, much longer than the timespan over which the famous iridium layer was deposited, and that the eruptions began about half a million years before the terminal Cretaceous event. During that time, ecosystems did not change dramatically, but they suddenly collapsed at the time of the Chicxulub event. In the light of all the data, the team concluded that a large asteroid impact 65 million years ago in modern-day Mexico was the major cause of the mass extinctions. One of the authors of the *Science* paper, Joanna Morgan, of Imperial College in London, put it like this:

We now have great confidence that an asteroid was the cause of the [terminal Cretaceous] extinction. This triggered large-scale fires, earthquakes measuring more than 10 on the Richter scale, and continental landslides which created tsunamis. However, the final nail in the coffin of the dinosaurs happened when blasted material was ejected into the atmosphere. This shrouded the planet in darkness and caused a global winter, killing off many species that couldn't adapt.

This brief summary, dramatic though it is, barely hints at the hazards faced by life on Earth in the aftermath of the impact. In view of its importance for our own existence, perhaps it is worth going into a little more detail.

THE ARCHETYPAL IMPACT

The shallow sea of what is now the Yucatán Peninsula provided an ideal location for the impact to do its worst. The sediments in that part of the world included carbonates and a thick layer of material known as anhydrite, rich in sulphur. So large quantities of both carbon dioxide and sulphur dioxide would have been produced in the impact, with the sulphur dioxide reacting with superheated water to produce huge amounts of sulphuric acid, carried up into the stratosphere in the form of tiny droplets. Immediately after the impact, before this acid haze could do its worst, the first major problem for

WHAT'S SO SPECIAL ABOUT THE CAMBRIAN EXPLOSION? (II)

life would have been heat and fire; the earthquakes and tsunamis barely rate a mention alongside the real catastrophes that were about to befall. Hot material thrown out by the impact would have travelled right around the globe, heating the entire surface with the equivalent of 10 kilowatts per square metre for several hours – an effect the American Jay Melosh has graphically described as 'comparable to that of a domestic oven set to "broil"'. This was enough to cause a global conflagration as the vegetation on land burnt, leaving a layer of soot still visible today alongside the iridium layer. The amount of soot present in the layer implies that 70 billion tonnes of carbon, equivalent to 25 per cent of all the organic material on Earth at the time, went up in smoke.

The darkness produced by all this smoke was enhanced by the presence of the sulphuric acid droplets in the stratosphere, which are very efficient at reflecting away the heat from the Sun. So photosynthesis would have been severely restricted, causing many plants to die, while in the aftermath of the 'broiling' the world was plunged into the grip of the global winter mentioned by Morgan. As the smoke cleared and the sulphuric acid rained out of the atmosphere, the world warmed again; but now, driven by the presence of all the carbon dioxide released in the impact, it overheated in a pronounced greenhouse effect. At the same time, the acid rain was bad news for any remaining life on the surface of the Earth, and especially bad news for sea creatures such as ammonites, whose shells were eaten away by the acid. And the ozone layer of the atmosphere must have been disrupted by all this activity allowing damaging ultraviolet radiation from the Sun to penetrate to the surface of the Earth. It has even been suggested that the ongoing volcanism on the other side of the world may have been stimulated into enhanced activity by shock waves spreading out from the impact site and being focused by the curvature of the Earth.

It is the sheer diversity of these environmental stresses that made the impact such a killer. Before this explanation came along, the experts were pushed to explain, for example, how one disaster could affect creatures as different as dinosaurs and ammonites, while leaving crocodiles, say, unscathed (or only a bit scathed). The answer is that there was a whole range of disasters, all triggered by the same event. On land, plant-eating dinosaurs suffered (as did their predators) when the plants died; in the sea, shelled creatures suffered because of acid rain. Survivors tended to be small (like our ancestors) so they could hide from the hazards and didn't need much food, or lucky enough to live in protected places from which they could spread out when the world began to get back to normal. Seventy per cent of species died; but that meant that 30 per cent had new opportunities to explore. But even this was small beer compared with the events that occurred at the time of Snowball Earth. A terrestrial impact cannot trigger such an event. As the events surrounding the death of the dinosaurs show, on the relevant timescales it is more likely to cause global warming than global cooling. But extraterrestrial events are another story.

COSMIC CLOUDS AND COMET DUST

If you don't put any numbers into the calculation, it's easy to imagine that if the Solar System passed through a dense cloud of interstellar material, the cloud would block out some of the heat from the Sun and cool the Earth, plunging it into a deep freeze. But even a dense interstellar cloud contains only about a million atoms and molecules, mostly of hydrogen, in every cubic centimetre. By most terrestrial standards that near enough counts as a vacuum – each cubic centimetre of the air that you breath contains about 40 billion billion molecules. And only about 1 per cent of the mass of the cloud would be in the form of solid particles, each about a tenth of a micrometre across. The sunshield effect would be far too small to affect the climate of the Earth at all.

There's just one loophole – if the Solar System happened to pass through a really dense cloud of material thrown out by a nearby supernova explosion. Then, dust high in the atmosphere of the Earth could indeed reflect sufficient solar heat to plunge the planet into a snowball state. But that would also leave traces of radioactive material in rocks of the appropriate age, and no such traces have been found.

WHAT'S SO SPECIAL ABOUT THE CAMBRIAN EXPLOSION? (II)

There is, though, a way in which the passage of the Solar System through an ordinary interstellar cloud could create an ordinary Ice Age, but not a Snowball Earth event. Hydrogen from the cloud would get captured by the Earth, and filter down into the upper reaches of the atmosphere. There, the molecules would be broken apart and the hydrogen atoms combined with oxygen to form a variety of compounds, including water vapour. Several people have considered the possible effects of all this on the stratosphere, where the ozone layer that shields us from solar ultraviolet radiation might be disrupted; but the most important effect would be to produce a dense, high-altitude cloud of water and ice crystals, reflecting away some of the Sun's energy and cooling the surface of the Earth. The maximum effect of such a tenuous cloud would, though, only be to reduce the surface temperature by about one degree Celsius, and that would only trigger an Ice Age if other factors were already tilting the climatic balance the same way. An interstellar cloud could not produce a Snowball Earth; but what about an interplanetary cloud?

An asteroid 10 km across striking the Earth was enough to trigger the terminal Cretaceous event. Such objects are far from rare, but the only ones that can do the damage are those that are in orbits that cross the orbit of the Earth. These come in two families: Apollo asteroids, named after the first of their kind to be discovered, spend most of the time farther out from the Sun than we are, but cross the Earth's orbit when they come close to the Sun; Aten asteroids, also named after their archetype, spend most of the time closer to the Sun than we are but cross the Earth's orbit when they are far from the Sun. A third group, the Amor asteroids, come close to the Earth's orbit (on the outside) but do not actually cross it. A few dozen Apollos are known (the number increases as more are being discovered), a similar number of Atens, and well over a thousand Amors, which are easy to spot because they are never far away. But the Apollos and Atens can only be seen when they are relatively close to us, and the statistics suggest that there are also thousands of them with diameters bigger than a kilometre.

Just how big could an Earth-crossing object be? The asteroid Eros is a roughly brick-shaped object, 35 km long, 11 km wide and 11 km deep. Think how much damage that could do if it hit the Earth. It is the archetypal Amor asteroid, and computer simulations suggest that its orbit may evolve into one that crosses that of the Earth in about 2 million years time. But as well as its long-term families of asteroids, and in spite of the shielding influence of Jupiter, the inner part of the Solar System still has visitors from farther out, beyond the orbits of the planets. These are the comets, lumps of ice with rocks embedded in them that originate in the Oort Cloud. Indeed, it is highly likely, although not absolutely proven, that many of the Earth-crossing asteroids are simply the rocky remnants of comets, with all the ice long since evaporated away, that plunged in from the outer Solar System long ago. And if Eros is a fragment, how big must the parent object have been?

There are two ways to tackle that question. First, we can look at how big icy objects in the outer Solar System are today. We can't see such objects as far away as the Oort Cloud, but we can see comparable objects in the Kuiper Belt. Pluto is best regarded as a Kuiper Belt Object rather than as a true planet; it has a diameter of roughly 2,300 km, and it isn't even the biggest Kuiper Belt Object. Could there really be comets that big? The other way to tackle the question is to try to reconstruct (in a computer model) one or more of the large objects that definitely did break up in the inner Solar System not so long ago. The best example is the trail of debris associated with Encke's Comet, named after the nineteenth-century astronomer Johann Encke, who was the first person to calculate its orbit. It moves between 0.34 astronomical units (AU) and 4.08 AU from the Sun, nearly getting as far out as Jupiter, in an orbit with a period of 3.3 years. This means it is the only active comet in an Apollo orbit, and regularly crosses the Earth's orbit. In the twentieth century, Victor Clube, of the University of Oxford, and Bill Napier, of the Royal Observatory, Edinburgh, joined forces to work out how Encke's Comet is related to a mass of other debris in a trail around the Sun.

When a dying comet settles into an orbit like this around the Sun, the icy material that holds it together gets eaten away each time it passes close to the Sun, literally disappearing in a puff of gas. That gradually releases the solid material that has been held in the grip of

WHAT'S SO SPECIAL ABOUT THE CAMBRIAN EXPLOSION? (II)

the ice since the birth of the Solar System, everything from asteroid-sized rocks down to particles the size of grains of sand. When the Earth crosses the orbit of this stream of particles, hopefully without encountering one of the larger rocks, the sand-grain sized pieces burn up in the atmosphere of our planet as meteors. This happens at the same time each year, when we cross the trail of debris, and so on the appropriate night, or over several nights, a shower of meteors seems to be coming at us from a point on the sky. These meteor showers are named after the constellations from which they seem to be coming. One of these showers, the Taurids, which seems to come from the constellation Taurus, arrives early in November each year; when the Earth is on the other side of the Sun, in late June, it passes through more debris in the same ring around the Sun, producing what are known as the Beta Taurids. This is a sign of just how spread out the debris trail is, and just how much material is in it.

In 1940, the pioneering comet researcher Fred Whipple worked out that the Taurid debris streams are in orbits exactly like that of Encke's Comet, but shifted sideways. He discovered that this is due to the gravitational influence of Jupiter, spreading the debris from the comet, and that it must have taken at least a thousand years for the amount of spreading we see to have taken place. Clube and Napier took the calculation further, finding that at least seven of the largest Apollo asteroids are associated with the Taurid streams, and that the largest of these, Hephaistos (which is about 10 km across), is in an orbit which suggests that it split off from the main body of Encke's Comet 20,000 years ago. Altogether, they estimated that there must be at least 150 major pieces of debris, each larger than a kilometre in diameter, associated with the Taurid streams and Encke's Comet. As they put it in their book *The Cosmic Winter*, 'it seems clear that we are looking at debris from the break-up of an extremely large object.'

Adding up all the material in the trails of debris, Clube and Napier estimated that the original object must have been at least a hundred kilometres across. This is about the size of Chiron, an icy object which travels around the Sun between the orbits of Jupiter and Uranus, crossing the orbit of Saturn. We know that many such objects, some many hundreds of kilometres across, exist in the outer Solar System, and the Encke's Comet/Taurid connection is proof that they can penetrate to the inner Solar System. What damage can they do when they get here?

DIAMOND DUST AND A FACELIFT FOR A GODDESS

Clube and Napier were especially interested in the effect on Earth of comets breaking up in the inner Solar System during historical and just pre-historical times. They investigated the way in which dust from such events spreads around the Sun and reflects sunlight to produce a phenomenon known as zodiacal light, which has changed in intensity over the centuries and millennia. They related these changes to climatic change on Earth, and suggested that an object like Chiron falling into an orbit like that of Encke's Comet might trigger an Ice Age. This freezing is actually quite hard to do; but the astronomer Fred Hoyle, known for his lateral thinking and more often right than wrong, suggested a way in which such events could tip the Earth into a deep freeze even more extreme than anything Clube and Napier discussed.

The nub of Hoyle's argument is that the climate balance of the Earth today rests on a knife edge, and only a relatively modest tilt of that balance could be sufficient to trigger feedbacks which produce a runaway effect. Climate has been likened to a 'weather machine', driven by the heat energy from the Sun. Water vapour is produced when solar energy is absorbed by the sea, and when that water vapour condenses back into liquid water in some other part of the globe that heat energy is released. So water vapour carries energy from the tropics to higher latitudes. It is also carried high into the atmosphere, to the top of the troposphere, about 15 km above the surface of the Earth. The temperature there is as low as -20 °C, but the water vapour doesn't all turn to ice, because it can only form ice crystals when there are tiny particles of dust or other material which act as 'seeds' on which they can grow. Without these condensation nuclei, the water stays as a supercooled vapour until the temperature gets down to -40 °C, at which point it is suddenly converted into a profusion of tiny ice crystals, which then act as the condensation nuclei for more water vapour to freeze, forming a mass of particles so reflective that they are known to Antarctic explorers as 'diamond dust'. This material is so reflective that if you took a layer of water equivalent to a thickness of one hundredth of a millimetre over the entire surface of the Earth and turned it into diamond dust crystals, each about one millionth of a metre across, it would reflect back into space almost all of the incoming solar energy. Turning just one tenth of 1 per cent of the water vapour in the Earth's atmosphere into diamond dust would be ample to trigger a Snowball Earth event – although Hoyle actually came up with his idea before the evidence for such events was discovered.

Hoyle was interested in finding a trigger for ordinary Ice Ages, not for Snowball Earth events. He developed a rather complicated model involving meteorite impacts which throw dust high into the air where it could cause sufficient cooling to trigger the formation of diamond dust. It is very difficult to do this without a huge impact, one which would have left a clearly visible crater on the surface of the Earth. But what if the impact didn't happen on Earth at all?

Although there have been many extinctions of life on Earth, the only ones that have been confidently identified with impacts are the two major events at the end of the Cretaceous and at the end of the Triassic. This isn't really good news, since it clearly implies that impacts are responsible for at least some of the worst disasters to strike the Earth. One way of assessing the risk is to study the surface of Venus, where in the absence of tectonic activity a cratering record has been left on the surface. This can be used to calculate the chance of an impact of a certain size occurring in a certain time interval – and since Earth and Venus are nearly the same size and are neighbours in the inner Solar System, essentially the same statistics apply to our home planet. Venus has been mapped by orbiting probes using radar, and its surface features are very well known. But, as I have mentioned, there is something odd about that surface. Compared with the Moon and Mercury, the density of craters is very low. Venus has too few craters for its size and age.

Studies of the Moon, Mercury and Mars tell us that Venus ought to be receiving an impact big enough to make a feature that shows up on the radar mapping about once every 700,000 years. There are roughly 900 such craters spread out at random on the surface, implying that the surface is between 600 million and 700 million years old. But the Solar System, presumably including Venus, is more than 4 billion years old. The age of the surface is only 15 per cent of the age of the planet. Traces of some recent volcanic activity have been observed on Venus – 'recent' being within the past few million years – but nowhere near enough to explain why the whole surface is so smooth. The explanation that planetary scientists have come up with is that about 700 million years ago (give or take 50 million years) there was a cataclysmic episode of volcanism on Venus, when the crust of the planet cracked and magma poured out, flooding across the surface, filling in old craters, and providing a blank slate on which new meteorite impacts could begin to leave their mark. Although catastrophic by geological standards, this resurfacing could have taken as long as a hundred million years.

This ties in with the evidence that Venus has a thick crust, and does not experience the kind of tectonic activity associated with continental drift on Earth. The usual interpretation is that beneath the insulating blanket of that thick crust, heat released by radioactivity builds up, increasing the pressure until something has to give and the crust cracks. On that scenario, Venus may have been resurfaced many times in its four-billion-year history.

But there is something else odd about Venus. It rotates the 'wrong' way, as we have seen. If you imagine looking down on the Solar System from a point in space far above the Earth's North Pole, you would see our planet rotating in an anticlockwise sense. This is the way all the planets rotate, except for Uranus, which is almost lying on its side, and Venus, which rotates backwards – clockwise, as seen from our imaginary vantage point. So the Sun rises in the west and sets in the east on Venus. But it takes a long time to do so. Venus rotates very slowly, once every 243 Earth days measured by the stars. This is longer than a year on Venus (225 Earth days), but as the planet moves around the Sun the length of a day, from noon to noon, is only 117 Earth days. So there are just under two Venusian days in each Venusian year.

WHAT'S SO SPECIAL ABOUT THE CAMBRIAN EXPLOSION? (II)

The explanation for the anticlockwise rotation of most of the planets is that it was built in by the direction in which the cloud of material from which they formed was orbiting the Sun. Uranus and Venus must have started out rotating in the same way – the laws of physics gave them no choice. But in each case their present pattern of behaviour can be explained if they were each hit a massive blow by an impacting object.

Nobody can say exactly when Venus received this blow, except that it must have been more than 700 million years ago or the traces of the impact would still be visible. Or rather, such an impact would have melted through the crust of Venus, releasing huge amounts of magma, flooding the surface and leaving no trace at all. James Kasting, of Pennsylvania State University, points out that the asteroid Ceres has a diameter of more than 1,000 km, and Vesta and Pallas each check in with diameters around 500 km. The impact of such an object with our planet, he says, 'would likely be enough to vaporize Earth's oceans entirely and create a steam atmosphere much like the runaway greenhouse atmosphere on early Venus ... [it] might sterilize the entire Earth.' But he doesn't ask, what if it hit Venus?

Although the evidence is only circumstantial, the temptation to draw the obvious conclusion is irresistible, unless you believe in stretching coincidence to the limit. If a really large object, the size of one of the objects in the Kuiper Belt, or a large asteroid, fell in to the inner Solar System about 700 million years ago and broke up, with the largest piece slamming into Venus, it could explain why Venus rotates backwards, why it was resurfaced at that time, and why the Earth, its upper atmosphere seeded with dust from the comet, briefly shone like a diamond as it reflected sunlight away from the upper atmosphere while the surface froze. Three puzzles solved by one supercomet. The fates of Hothouse Venus and Snowball Earth may have been inextricably linked as the scene was set for the Cambrian explosion. And the rarity of such an event merely highlights how special the Cambrian explosion was, and how lucky we are to be here.

So – what was it that made our species special?

8
What's So Special about Us?

The obvious thing that makes us special is our intelligence – more specifically, bearing in mind the dolphin and others, our kind of intelligence – even if it is rather hard to define exactly what we mean by intelligence. Intelligence is not solely a function of brain size, as any woman will tell you, but it is clear that in some sense dolphins are intelligent, and they have large brains in proportion to their body size – in some species, larger in this sense than the brain of the chimp or the gorilla. Indeed, until about 1.5 million years ago, dolphins were the largest-brained creatures on Earth by this measure, and therefore, perhaps, the most intelligent. It was only then that the brain of *Homo erectus* developed to overtake that of the dolphins. But it was the *Homo* line that discovered, or invented, technology, not (partly for the obvious environmental reasons) the dolphin line.

To paraphrase Saint Augustine: What then is intelligence? If no one asks me, I know what it is. If I wish to explain it to him who asks, I do not know. What we do know is that human intelligence seems to have been a unique product of evolution on planet Earth. And yet, there is some evidence that something very nearly like it might have emerged sooner, given the chance. This is perhaps the most important example, from our perspective, of the interplay between contingency and convergence – or as the Nobel Prize-winning French biologist Jacques Monod put it, rather more elegantly, chance and necessity.

WHAT'S SO SPECIAL ABOUT US?

CHANCE, NECESSITY AND THE DECIMAL SYSTEM

In this terminology, it is chance that decides which species survive a catastrophe such as the terminal Cretaceous event, and necessity that determines how the survivors evolve as they adapt to the conditions in which they live. If being an intelligent, upright biped with stereoscopic vision and hands with which to manipulate things confers such a huge evolutionary advantage today, why didn't the pressures of natural selection, operating over more than 150 million years while the dinosaurs dominated the Earth, produce an intelligent, upright dinosaur?

One answer is that during that long interval of geological time the evolutionary pressures were, in many ways, less pressing than they have been in more recent times. It was a time, by and large, of environmental stability. In particular, because of the geographical arrangement of the continents, there were no great Ice Ages to weed out species and put a premium on intelligence and adaptability. More of this shortly. But the other answer to that question is that although the mills of evolution ground slowly, they did still grind, and towards the end of the Cretaceous a very interesting kind of dinosaur did emerge.

Dinosaurs weren't just great lumbering brutes with tiny brains. The term 'dinosaur' covers as wide a variety of creatures as the term 'mammal' does today. There were the equivalent of carnivores such as lions and tigers, the equivalent of grazers such as deer and sheep, and even aquatic and flying relatives (not, strictly speaking, dinosaurs). Among this variety, one family that provides particularly relevant insight into the reason why we are here emerged. The archetypal member of that family is a species known as Troodon, which gives its name to the genus and also to the family itself, the Troodontidae. It used to be known as Stenonychosaurus, and had a close relative called Saurornithoides; in the present context, for non-specialists all three names are equivalent, but I shall stick to Troodon (which is pronounced with three syllables, Tro-odon).

Troodon was a relatively small dinosaur, about 2 or 3 metres long from head to tail, which stood on two legs and had two arms, each ending in three slender digits, one of which was partly opposable, like our thumbs. It weighed about 50–60 kilos and had large eyes (which some experts believe implies that it was nocturnal), and the eyes were placed towards the front of its head, so it had stereoscopic vision. Its teeth suggest that it was omnivorous, although primarily a carnivore. It was a lightly built, agile hunter with good vision and good grasping hands; in all probability its prey included not only small reptiles but small mammals, including our ancestral species. Most significantly, it had the largest brain in proportion to its body mass of any dinosaur, with a ratio not far short of that of a modern baboon. All these signs suggest that Troodon was well on the way to developing our kind of intelligence. But it had the misfortune to live right at the end of the Cretaceous period.

The terminal Cretaceous event killed off all land animals bigger than about 40 kilos in mass, including Troodon itself and all the other members of its genus and family. But what if the disaster had never happened? This tantalizing scenario has encouraged several scientists, notably the palaeontologist Dale Russell and the astronomer Carl Sagan, to try to extrapolate the future evolution of Troodon if the terminal Cretaceous event had never happened. Russell and his colleague Ron Seguin went so far as to construct a 'life size' model of what they called a 'dinosauroid', a large-brained, reptilian biped with large eyes and three-fingered hands with opposable thumbs.

Russell calculated that at the rate of evolutionary change that was going on in the late Cretaceous, a creature with a body mass the same as that of a modern human being, and with a brain to match, would have taken about 25 million years to evolve, emerging around 40 million years ago. Such an intelligent dinosauroid would, it seems likely, have developed an arithmetic system built upon base six, for the same seemingly logical (but actually arbitrary) reason that we use base ten. In 40 million years of further evolution, what might such a species have achieved? As Sagan rather conservatively put it (writing before the cause of the terminal Cretaceous event was known):

If the dinosaurs had not all been mysteriously extinguished some 65 million years ago, would the Saurornithoides have continued to evolve into increasingly intelligent forms? Would they have learned to hunt large mammals collectively and thus perhaps have prevented the great proliferation of mammals that followed the end of the Mesozoic Age? If it had not been for the extinction of the dinosaurs, would the dominant life forms on Earth today be descendants of Saurornithoides, writing and reading books, speculating on what would have happened had the mammals prevailed?

Enthusiasts for SETI argue that if Troodon came so close to evolving intelligence so long ago, this increases the likelihood of intelligence being found elsewhere in the Galaxy. But the Fermi paradox returns with renewed force, since the puzzle is no longer just why civilizations from other planets have not left their calling cards in the Solar System, but why there have been no previous spacefaring civilizations on Earth to leave traces in Earth orbit, or on the Moon, or on Mars. The pessimistic point of view is that it took 150 million years of dinosaur evolution to get on to even the first rung of the ladder leading to our kind of intelligence, and then disaster struck. Given how often disaster has struck life on Earth, how could intelligence ever have time to evolve? In our case at least, the answer seems to be that climatic convulsions cranked up the pace of evolution.

THE MOLECULAR CLOCK

We know how quickly *Homo sapiens* evolved from the mammalian equivalent of something like Troodon from a combination of fossils and molecular evidence. DNA 'fingerprinting' is now a familiar technique from its use in forensic investigations. DNA from a sibling, for example, is routinely used to identify the victims of disasters in which human remains cannot be identified by other means. Siblings, or parents and their children, have very similar DNA. The DNA of cousins is slightly more different than that of siblings, more distant relations have even more differences in their DNA, and so on. If you had a group of people of different ages, all related to one another, it would be easy, using measurements of their DNA alone, to work out a family tree for the group, perhaps showing that they were all descended from the oldest person present, their common ancestor. You can do the same thing for species. By analysing their DNA, you can work out which species are closely related (like siblings), which ones are more distantly related (like cousins), and so on. As in human families, sibling species have recent common ancestors (the equivalent of parents), cousin species have more distant common ancestors (the equivalent of grandparents), and so on.

It was this kind of analysis which showed, for example, that the giant panda is a bear – until the DNA technique became available, zoologists had been uncertain whether to classify it as a kind of bear or a kind of raccoon. Similar studies resolved a debate among the experts about the ancestry of the dog. Prior to the use of DNA, the experts were divided into two schools of thought. One thought that early dogs were descendants solely of tamed wolves; the other suspected that jackals or coyotes also contributed to the dog's ancestry. The DNA evidence rules out any ancestor species except the wolf. Such studies have established many other relationships that had proved difficult to assign on anatomical evidence alone. But what of our own species?

It has become something of a cliché, but it is none the less true, to say that we share 99 per cent of our genetic material with chimpanzees. More precisely, our common genetic heritage amounts to about 98.6 per cent of our DNA, but that is still remarkable in view of the superficial differences between ourselves and chimps. There are actually two species of chimpanzee, the common chimp (*Pan troglodytes*) and the pygmy chimp (*Pan paniscus*). The DNA technique is sufficiently sensitive to tell us that the pygmy chimp is our closest living relative, and that the common chimp is a slightly more distant relative. But only slightly. By all the usual rules of biology, we should be classified as chimpanzees too, as *Pan sapiens* – it is only our natural inclination to see ourselves as something special that leads us to be classified in a separate genus, *Homo*, all on our own. Our next nearest living relatives are, as you might expect, the gorilla and the orang-utan. But if these are our cousins, so to speak, who were our grandparents, the common ancestors of ourselves, chimps, gorillas and orangs, and when were they alive? Putting it another way, how fast does the molecular clock tick?

The first step is to establish that the clock does tick at the same rate in all the species we are interested in – especially ourselves, the chimps and the gorillas. The changes in DNA have accumulated through mutations which have taken place since the split from a common ancestor, and it would be wrong simply to assume that mutations occur at the same rate in each species, no matter how plausible such an assumption might seem. The molecular clock (or clocks; other molecules than DNA can also be used) has to be tested for accuracy and calibrated against other evidence such as fossils before being used to establish family trees.

In practice, this is a long process involving detailed investigations of many species, their molecules and other evidence. These details need not concern us here, but one example will give a flavour of how the technique works. The apes (including ourselves), baboon and squirrel monkey share a common ancestor back in the evolutionary past – just how far back doesn't matter at this stage. The molecular 'distance' between ourselves and the squirrel monkey is 15 per cent, and this is exactly the same as the molecular distance between the squirrel monkey and the baboon, and between the squirrel monkey and the other apes. So the same amount of change has accumulated in the same amount of time in all these species. This shows that the molecular clock has been ticking at the same rate in all these species, ever since the split. In a similar way, the distance between the apes and the baboon is the same for all the apes. So we know that the clock has been ticking at the same rate in humans and the other apes at least since the time of the split from the common ancestor with the squirrel monkey. This is further evidence that in evolutionary terms there is nothing special about the human line; we are just a variety of African ape, evolving in the same way that other apes evolve.

Although there are no fossils which can be used to date precisely the split between the human line and our sibling apes, there are fossils which can provide dates for earlier splits, and we can use these to calibrate the clock. The bottom line is that the combination of fossils and molecular evidence tells us that the split between humans and chimpanzees occurred a little less than 4 million years ago, and that the gorilla line split from our ancestral line just before the human–chimp split.

The question this raises is, what happened to cause the split? What happened to cause a population of forest-dwelling apes to split into several lines, most of which stayed in the forests of Africa, but one of which set out on a course that would lead it to become the dominant species on Earth, building radio telescopes and spaceprobes and puzzling over the possibility of intelligent life elsewhere in the Universe? We know, from fossil evidence, where the split occurred, in the region of the African Great Rift Valley; we know when it occurred; and there is an outstanding candidate for the reason why it occurred, and why it happened so quickly, compared with the pace of evolution at the end of the Cretaceous.

THE TRIGGER FOR CHANGE

What makes us different from the other species of African ape? Adaptability and intelligence. What was the advantage of adaptability and intelligence to an African woodland ape a little more than 4 million years ago? A series of environmental changes that drastically affected the African forests, and had their origins in the shifting arrangements of the continents.

About 5 million years ago, conditions on Earth changed sufficiently significantly for geologists to set that date as the end of an epoch, the Miocene, and the beginning of a new epoch, the Pliocene (epochs are the subdivisions of geological periods). It was during the Pliocene that the first true hominids appeared, and the first species classified as *Homo*. There was very little geographical change going on in the far south of our planet at that time, except that Australia and South America were moving north, away from Antarctica, which was already established over the South Pole. This allowed a strong ocean current to circulate around Antarctica, cutting it off from warm tropical currents, so that around 6 million years ago it became so cold that even more ice was locked up in the southern ice cap than is there today. With so much water frozen, the global sea level was 50 metres lower than it is today. Among other things, this caused the Mediterranean Basin to dry out completely, not once but several times as the ice cap in the south varied in size.

With the south locked in to a fairly stable pattern with relatively minor fluctuations, it was the changing geography of the Northern Hemisphere that came to dominate the story of our origins. As the continents moved northwards, surrounding and closing in on the Arctic region, the environment changed for two reasons. First, with land farther to the north it was easier for snow to settle on it and build up into ice sheets and glaciers. Secondly, although the Arctic Ocean remained over the North Pole itself, it became increasingly landlocked, so that warm tropical currents could not penetrate into it, and it cooled. This eventually allowed an ice cap to form over the sea itself – and once an ice cap forms, its shiny surface reflects away incoming solar energy and helps to maintain cool conditions. As far as anyone can tell, this pattern, with a land-locked northern ocean covered in ice and simultaneously a continent over the South Pole covered in ice, has never previously occurred during the long history of our planet. It is as unique as we are.

The first effect of the poleward shift of the continents in the Northern Hemisphere was to produce a bigger variety of climatic environments. When all the land was nearer the equator, the climate over much of that land was essentially what we would call tropical. This situation, which had persisted for tens of millions of years, was very good for species that had adapted to such an environment, but didn't provide much opportunity for the kind of environmental pressures that stimulate evolution. Around the time the Mediterranean Sea was experiencing its bouts of desiccation, though, the Arctic region was experiencing a cool, temperate climate, with coniferous forests extending right up to the northern limits of the land. Although there may have been seasonal snows, the high northern latitudes were kept ice-free for a time as the Gulf Stream, which carries warm water up from the south, intensified. This happened as the gap between South America and North America slowly closed up between about 5 million and

3 million years ago, leaving the water with no escape route through to the Pacific. But as the path to the far north became blocked by Greenland, the current was deflected to the east, where it acts to keep northwest Europe warmer than Newfoundland today, and the Arctic eventually froze. By that time, the gap between the Iberian Peninsula and Africa had widened slightly as a result of tectonic activity, and the Mediterranean Basin had filled with water for the last time, with an inrush through the Strait of Gibraltar flowing at speeds of up to 300 km per hour completing the job in less than two years, about 5.3 million years ago. Sea level rose by as much as 10 metres per day.

With a bigger variety of climates, there was a greater variety of environments for both plant and animal life on Earth. When much of the land was covered by tropical forest, there was plenty of scope for tropical species but very little for anything else. When the world became divided into tropical, temperate and other zones this was bad news for many tropical species, at least those on the fringes of the forest, which had to compete with one another for diminishing resources; but it was good news for species able to move out of the forest and adapt to different lifestyles. So the variety of life on Earth actually increased after about 6 million years ago. The ancestors of modern dogs emerged at that time, quickly followed by, among others, the lines leading to modern bears, camels and pigs. And in East Africa, as we have seen, there was a three-way split in the ape line, with one of the offshoots from that split leading to the emergence of *Homo sapiens*.

THE PACEMAKER OF HUMAN EVOLUTION

The geological conditions in East Africa around 5 million years ago made the region particularly sensitive to the kinds of climatic changes the Earth was experiencing. At exactly the time of the three-way split along the line leading to the African apes, tectonic activity was lifting a slab of the Earth's crust and then cracking it apart by sideways forces to create the great East African Rift Valley system. Instead of being covered by tropical forest enjoying a year-round damp climate, the region became drier, and the extent of the forest became more limited. The climate became slightly seasonal, as part of the overall pattern of climatic changes going on, and islands of tropical forest were now surrounded by grasslands and regions of more open woodland.

As far as temperature was concerned, the region was not dramatically affected even when ice sheets spread farther to the north, although of course it did cool. What was far more important to the flora and fauna of the Great Rift Valley was the associated change in rainfall. When the world cools, there is less evaporation from the oceans, so there is less moisture in the air and less rainfall overall. And when ice caps grow, sea level falls and large areas of continental shelf are exposed, so that rain-bearing systems have farther to travel before reaching a region such as the Great Rift Valley, and dump much of their moisture before they get there. What matters, in terms of the evolution of our own ancestors, is that an Ice Age at high latitudes is a dry age in East Africa. And drought is bad for forests.

By 3 million years ago, the Arctic was nearly as cold as it is today, and although snow and ice had not spread far, there was a distinct cooling and drying in the equatorial regions, revealed by various kinds of geological evidence. Ice appeared on the continent of Europe and on the mountains of California by 2.5 million years ago. Today, the area of the Earth's surface covered by ice is about 15 million square kilometres; but by about 2 million years ago the area covered by ice was 45 million square kilometres, and the volume of water locked up in ice was, judging from the fall in sea level at that time, 56 million cubic kilometres. 'Coincidentally' (I do not believe it was merely a coincidence), around that time the first member of the *Homo* line to move out of Africa began to spread first throughout Africa and eventually into Europe and Asia. This was a species known as *Homo erectus*, which was all but human below the neck and had a brain capacity of about 900 cubic centimetres when it started to spread, evolving into a capacity of about 1,100 cubic centimetres as time passed; the modern human brain capacity is 1,360 cubic centimetres. Clearly, a large brain and all that goes with it – intelligence and adaptability –

was central to the success of *Homo erectus* and the later variations on the *Homo* line that it evolved into. The reason is easy to find in a regular pulsebeat of climatic change that developed around this time.

Although there are several cycles involved in a complex interaction of climate rhythms, the main pulsebeat, revealed by geological studies, is a cycle in which full Ice Ages, each roughly 100,000 years long, are separated by warmer intervals, known as Interglacials, each roughly 10,000 years long. We live in an Interglacial that began a bit more than 10,000 years ago; all of human civilization is contained within one climatic pulsebeat. But, of course, the 'next' Ice Age may not appear on schedule, because of the impact our technological civilization is having on the climate of the Earth.

This pattern is well understood. It has to do with the way the tilt and wobble of the Earth change slightly over the millennia (in spite of the stabilizing influence of the Moon) and the fact that the shape of the Earth's orbit changes slightly, from more circular to more elliptical and back again, due to the gravitational influence of the other planets, in particular Jupiter. The resulting theoretical description of how climate changes is often called the Milankovitch Model, after the Serbian astronomer Milutin Milankovitch, who worked out the details. There are actually three main 'Milankovitch' cycles, one roughly 100,000 years long, one about 43,000 years long, and one around 23,000 years long. The important point is that although the total amount of heat received by the Earth from the Sun over the course of a year stays the same, the way the pattern of heat is distributed through the seasons changes. And the unusual distribution of continents over the past few million years has left Antarctica in a permanent deep freeze but made the Northern Hemisphere highly sensitive to these changes. This is a unique situation in Earth history, and one responsible for our existence.

Sometimes, northern winters are very cold but summers are very hot; sometimes, winters are mild but summers are cool (the pattern is, of course, reversed in the Southern Hemisphere). With so much land at high latitudes around the frozen northern ocean today, the natural state of our planet is to be in the grip of a full Ice Age. According to the Milankovitch Model, the Earth can only be tugged out of a full Ice

Age for a few millennia when conditions are just right to produce hot norhern summers, warm enough to melt much of the ice. As long as the summers remain hot, once a little ice melts the process accelerates, because the dark land revealed as a result absorbs more heat from the Sun than the shiny ice did. But once the summers cool off, since it is always cold enough for snow to fall in winter the process goes into reverse, and the planet reverts to a full Ice Age. The geological record exactly matches this prediction.

Farther south, in the forests where our ancestors made a living, the result was a repeating pattern of drought and plenty. The permanent ice cover in the south already implied less moisture and made the forests more sensitive to the rhythms of the Milankovitch cycles. When times were hard, and the forests shrank, the proto-apes that lived in them were faced with two choices. They could retreat into the heartland of the forest, competing for scarce resources with other tree dwellers and evolving into more efficient tree apes, including the chimp and the gorilla, as a result. Or they could scrabble a living on the fringes of the forest, learning a new way of life and beginning to move out onto the plains. It was the less successful tree dwellers that were pushed to the margins and had to adapt or die.

If the drought had continued, they would probably all have died. Many of them did. But after something like 100,000 years the rains returned, the forest expanded, and the living was easier. The survivors from the winnowing would have been the toughest and most adaptable proto-apes, the fittest in the evolutionary sense, best suited to the new way of life. There would have been a population explosion, spreading the genes responsible for their success, before the drought returned and the screw was tightened once more. The pattern repeated for millions of years, and each turn of the screw first selected for intelligence and adaptability then provided a respite during which the survivors could thrive. No wonder our ancestors evolved more quickly than Troodon!

The pacemaker of Ice Ages was the pacemaker of human evolution. If the ice had come in full force 3 or 4 million years ago and stayed, East Africa might have become mostly desert, with only the best-adapted tree apes able to survive in the last vestiges of the forest. Conversely, if there had been no droughts there would have been no evolutionary pressure for the selection of our kind of intelligence – as is borne out by the example of South America, where the tropical forest survived and the tree-dwelling primates continued to do very well in their old role. Our ancestors survived and evolved, so we are here, because of the unusual pattern of climatic changes that actually happened. But they only just survived – the DNA evidence tells us that there is more genetic difference among individuals within a single chimpanzee troop in West Africa than among all living humans on Earth; it is estimated that all human beings alive on Earth today are descended from a population of less than a thousand early humans. That's how close the Earth came, even after surviving all the earlier hurdles, to not having a technological civilization. But it is no coincidence that, once our ancestors had reached a certain level of intelligence and adaptability, civilization emerged during an Interglacial, with populations booming in times of plenty, the development of agriculture, and all the rest. But what is our future?

THE FATE OF TECHNOLOGICAL CIVILIZATION

It is not possible to quantify all of the steps in the chain of circumstances that has led to our existence, no matter how much fans of the Drake equation might wish to do so. But it is clear that the probability of many, if not most, steps in the chain is small. Planets are common – but not Earth-like planets. Life is likely to emerge in the oceans of an Earth-like planet – but it is unlikely to evolve into complex multicellular creatures. And so on. You may feel that I err on the side of pessimism. But however optimistically you assess the odds of arriving at something like Troodon or a South American monkey, the final nail in the coffin of the hope that intelligent creatures like ourselves might be common in the Milky Way lies in the string of coincidences that, putting it simplistically, turned a monkey into a man. It required, simultaneously, ice over both the North and the South poles, a variety of climate zones, the geological changes that produced the East African

Rift system, and a planet with just enough wobble to produce the Milankovitch Ice Age rhythms! We are a very unlikely species.

This ought to give us a sense of responsibility. Of course, it hasn't. Lovelock's vision of the Earth as a single living system, Gaia, helps to explain why conditions suitable for complex life have been maintained for so long on our planet, but it also highlights how vulnerable the Earth System is to the kind of sudden shocks being imparted to it by our technological civilization. The big climatic concern today is global warming caused by human activities. Most of that concern is expressed in terms of the implications of a rise in global mean temperature of, say, two (or three, or four) degrees Celsius by the middle (or end) of the twenty-first century, with the implication that this rise will be smooth and gradual. But the geological record reveals many occasions when the Earth's temperature has changed abruptly, either up or down, when a tipping point is reached. The simplest way to picture this is to think of the ice cover of the Arctic Ocean. When the Arctic is covered by ice, it reflects away solar energy, and even if the amount of incoming radiation were to increase the temperature would only rise slowly. But once the ice began to melt, the dark ocean would absorb much more energy and there would be a sudden rise in temperature. When the amount of energy reaching the surface begins to fall, the feedback goes into reverse, with the ocean cooling only slowly to the point where ice forms, then a sudden fall in temperature. Lovelock calculates that other feedbacks, involving both living and non-living components of the Earth System, could make the temperature of our planet jump by 4–6 °C when a tipping point is reached by the middle of the present century. In geological terms, such a change happening over 10,000 years would be sudden, and would cause mass extinctions; we may be causing such a change over a human lifetime. Life would survive; whether technological civilization would survive is an open question.

Global warming is not the only threat posed by the buildup of carbon dioxide and other waste products from our activities. An equally important hazard, which is only just beginning to get the attention it deserves, is the acidification of the oceans which occurs when carbon dioxide from the atmosphere dissolves and reacts with water to make carbonic acid. The most obvious effect of this is to dissolve away coral reefs, but the acid attacks the shells of many marine creatures, including the tiny plankton at the bottom of the food chain. Extreme acidification could lead to widespread desertification of the oceans, with unimaginable consequences for the rest of life on Earth, and soaring temperatures as more carbon dioxide is released from the dying oceans.

In the short term, the biggest threat to technological civilization seems to be technological civilization – and that is without considering the possibility of war on a global scale. In the slightly longer term, the biggest threat seems to be the same one that finished off the dinosaurs – impact from space.

Such an impact doesn't have to be on the scale of the event that did for the dinosaurs, let alone on a scale sufficient to cause a Snowball Earth event, in order to be bad news for us. And the kinds of impact that would be bad news are by no means rare. A large part of the problem is that the myriad objects in the Asteroid Belt do not all stay quietly in orbits between Mars and Jupiter. With hundreds of thousands of asteroids now identified, and the ability to calculate their orbits both forwards and backwards in time, astronomers have discovered that the impact that caused the terminal Cretaceous event was a direct result of a collision between two big asteroids that occurred roughly 100 million years before the dinosaurs met their doom. This chance encounter between two objects 170 km and 60 km in diameter, meeting in a head-on collision at 11,000 km per second, splattered about 150,000 pieces of debris into new orbits where the gravitational influence of Jupiter spread much of it like a shotgun blast across the inner Solar System, where it pebble-dashed the inner planets and our Moon, with one of the pieces killing off the dinosaurs and another producing the young lunar crater Tycho.

On a lesser scale, just over a hundred years ago, in the summer of 1908, a piece of cosmic debris burnt up in the Earth's atmosphere and exploded over the Tunguska region of Siberia. The blast devastated a region of 2,000 square kilometres and felled 80 million trees, laying them out like matchsticks pointing away from the centre of the explosion. It is estimated that the incoming space rock was moving at about

15 km per second (over 50,000 km per hour) and reached a temperature of 25,000 °C before the heat blew it apart, at an altitude of about 10 km. The energy released was the equivalent of the explosion of a 10 megatonne nuclear bomb. But the rock that caused all the damage was only about 30 metres across. By chance, this happened over one of the most desolate, almost uninhabited, regions of the globe. If the Tunguska meteor had arrived on the same trajectory but just a few hours later, as the world would have turned a little on its axis the explosion would have occurred almost directly over St Petersburg, destroying the city and all its inhabitants.

Between these extremes – the death of a city and the death of the dinosaurs – something dramatic happened to the mammals a little less than 13,000 years ago. At that time, temperatures in some parts of the world fell by as much as 15 °C in a short time, and at least thirty-five mammal species, including the mammoths, went extinct. This time, people were affected: the Clovis civilization of North America also went the way of the mammoths. The most likely explanation for all this is that an object significantly larger than the Tunguska meteor exploded in the atmosphere above North America and rained debris on the land below. Dust and smoke from fires triggered by the event shrouded the globe and cooled the surface, in the same way, but on a much smaller scale, that it was cooled by the after-effects of the Late Cretaceous impact. What many experts regard as convincing proof that the disaster was caused by an impact has been found in the form of tiny diamonds, scarcely visible even under the microscope, in deposits of the right age across North America and Europe. These 'nanodiamonds' can only be produced by the extreme heat and pressure associated with an impact. The comet responsible for such a widespread disaster must have been about 5 km across, but had probably broken up into smaller pieces before hitting the Earth, rather like the way the comet Shoemaker-Levy 9 broke up before hitting Jupiter in 1994.

In the light of the evidence that such events are normal, and certain to happen again sooner or later, with the only comfort being that larger impacts are more likely to happen later, a serious (but rather modest) effort is now going in to the construction of telescopes that will identify and monitor the orbits of thousands of asteroids and comets that are in orbits bringing them uncomfortably close to the Earth (Near Earth Objects, or NEOs). NASA's WISE satellite (the acronym stands for Wide-field Infrared Survey Explorer), launched late in 2009, detected hundreds of 'new' asteroids every day during its first three months of operation, including five NEOs. But identifying these NEOs is one thing; doing something about it if we find one on a collision course with the Earth is another. The official estimate is that there is a 1 in 100 chance that a rock at least 140 metres in diameter, large enough to cause widespread damage across an entire US state, or along the Atlantic seaboard of Europe, will hit our planet within the next 50 years, and as things stand we could do nothing to stop it. Impact from space is clearly the biggest natural threat to civilization, and a plausible reason why we are alone, assuming similar impacts occur, as they surely must, in other planetary systems.

A less quantifiable, but no less real, threat is posed by volcanism. There have been devastating 'supervolcanoes' even in the relatively recent geological past, including the eruption that produced Lake Toba, in Indonesia, some 70,000 years ago. This was the largest known eruption of the past 25 million years. To give an idea of its size, it spread a layer of ash roughly 15 cm thick over the entire Indian subcontinent. Some researchers have suggested that the environmental impact of the eruption almost wiped out the human populations of Central Eastern Africa and India at that time. Today, the entire region under Yellowstone Park in the United States has been identified as just such a supervolcano, in a quiescent state just now but which could explode at any time, with very little warning. And on an even larger scale, events like the one in which the Deccan Traps formed can surely happen again. Whichever way you look at it, with disaster striking either from within or from without, our civilization is doomed, and the only real question is when the end will come. There has been time between disasters for a technological civilization to emerge on Earth, but that is no guarantee that a similar opportunity will have arisen anywhere else in the Milky Way. Even here, though, our civilization may only just have emerged in time before the Earth ceases to be a comfortable home for life.

THE FATE OF THE EARTH

The ultimate fate of the Earth is to be burned to a cinder when the Sun swells up to become a red giant star as it approaches the end of its life. Exactly when this will happen is partly a matter of conjecture. As the Sun ages, it loses mass into space, so its gravitational pull will weaken, and the Earth's orbit (along with the orbits of all the other planets) will expand, perhaps delaying the inevitable. But the tenuous gas in the extended atmosphere of the ageing Sun will drag on the Earth by friction, probably sending it spiralling in to meet its doom sooner than if it stayed in the same orbit and waited for the Sun to engulf it. Either way, though, the end will come very roughly in about 5 billion years from now, since the Sun is roughly halfway through its time as a Main Sequence star.

At first sight, this looks like good news for people hoping to find other technological civilizations in our Galaxy. It looks as if, even if we had to start again from scratch, there would be time for a second such civilization to develop on Earth, and therefore – presumably – on other Earth-like planets orbiting Sun-like stars. The chance of a civilization like ours arising seems to have doubled. But think again. If complex life on Earth were destroyed tomorrow, the prospects would not be the same as they were in the Precambrian. The Earth is likely to lose much of its air and all of its water, becoming a hot desert planet, long before it is engulfed by the Sun. Particles of gas escape from the top of the atmosphere all the time, but only if they are travelling faster than the escape velocity from the Earth, which is just under 11 km per second at that altitude. Very few particles travel at that speed today; but a trickle of hydrogen atoms, produced by the breakdown of water molecules in the upper atmosphere, does escape. That trickle will become a flood in about a billion years from now, because as the Sun ages it is getting brighter by about 10 per cent every billion years. A brighter Sun will make the Earth warmer, evaporating more water from the oceans and enriching the upper atmosphere with water molecules that can release hydrogen into space. Three billion years from now, all the water will have gone.

But this astronomical calculation assumes nothing else is changing. When other factors are taken into account, it may not even take that long to turn the Earth into a lifeless world. Just as the Sun will be hotter when it is older, so it was cooler when it was younger. In spite of this, the young Earth was kept warm enough for liquid water to exist by the greenhouse effect of gases such as carbon dioxide in its atmosphere. But as the Sun has warmed, Gaian mechanisms have reduced the amount of these gases in the atmosphere, so that the temperature on Earth has remained in the range where liquid water can exist. Lovelock has pointed out that this process cannot continue much longer, because there is very little carbon dioxide left in the atmosphere to remove. Even if we ignore the blip caused by human activities, if our technological civilization did not exist, the Earth would still overheat and lose all its water, but on a timescale of hundreds of millions, rather than billions of years.

Still, a hundred million years is a long time. Longer than the time that has elapsed since the death of the dinosaurs. Plenty of time for other civilizations to develop if ours fails. Unfortunately, it seems that Earth will be lucky to survive much more than a million years as a home for life.

In his book *Extinction*, David Raup, of the University of Chicago, looked at the pattern of mass extinctions throughout Earth history. One of the fruits of this research is that it is possible to measure the rate at which extinctions of different sizes occur, and therefore to estimate how long it is before the next extinction of a particular size is due. Out of curiosity, Raup pushed this technique to the limit to estimate how big an interval there should be between extinctions large enough to destroy all life on Earth. He seems surprisingly sanguine about the number he came up with:

I once tried extreme-value statistics on extinction data to ask, 'How often should we expect extinction of all species on Earth?' I don't have much confidence in the results, but they are at least comforting. Extinctions sufficient to exterminate all life should have an average spacing of well over 2 billion years.

Comforting? Perhaps, if you start counting the 2 billion years from now. But not if you remember that there has been life on Earth for nearly 4 billion years. According to the statistics, we are well overdue for the end.

As the old saw has it, 'There are lies, damned lies, and statistics.' Nobody seems to have bothered much about Raup's calculation. And then, in 2010 came evidence that the statistics might be telling us something worth paying attention to after all.

Vadim Bobylev, of the Pulkovo Astronomical Observatory in St Petersburg, was analysing data from a European satellite called Hipparcos when he discovered that a nearby star, known as Gliese 710, is on a collision course with our Solar System. Gliese 710 has about half as much mass as our Sun, and is now about 63 light years away from us in the direction of the constellation of the Serpent. It is heading our way at a speed of roughly 50,000 km per hour, and is destined to pass through the Oort Cloud of comets at the fringes of the Solar System within a million and a half years; it might even get as close to the Sun as the Kuiper Belt. The disruption of the Oort Cloud resulting from this close encounter will send debris into the inner Solar System on a scale not seen since the Late Heavy Bombardment. If Gliese 710 has its own comet cloud, as seems likely, the bombardment will be even more intense. Without a doubt, this will result in the extermination of all life on Earth, and send our planet back to the kind of state it was in just after the formation of the Moon. That's how close the Earth came to never having a technological civilization – a million years, after nearly 4 billion years of evolution.

But still, that gives us a million or so years to play with. And if our civilization did collapse, although there might not be time for another intelligent species to evolve, there would be time for any human survivors to build a new technological civilization, wouldn't there? Unfortunately, there's a snag. There might be time, but they wouldn't have the resources.

NO SECOND CHANCE

Judging by the example of life on Earth, the development of technological civilization on any sort of large scale might not be possible until large reserves of fossil fuel have built up in the crust of a planet. It is no coincidence that our kind of civilization developed first in a part of the world where there were easily accessible supplies of coal near the surface of the Earth, and ores such as iron and copper that could be mined with equipment no more advanced than a pick made from the antlers of a deer. But it took billions of years of tectonic activity to lay down those ore deposits, distort the strata to bring them near to the surface, and for erosion to do its work of exposing them. Even the coal deposits on which our technological civilization was almost literally built were largely laid down between about 360 and 280 million years ago, in the cool, swampy environment of the appropriately named Carboniferous period. If those environmental conditions had not persisted for tens of millions of years, could a technological civilization ever have arisen on Earth? The later Carboniferous also saw the formation of important oil deposits. Easily accessible oil fuelled the dramatic growth of technological civilization in the twentieth century.

But all these resources have been depleted dramatically. We can still extract mineral ores, coal and oil in large quantities, but only by using technology, metals, and fossil fuel to do the job. There is no better example than the offshore drilling rigs of the North Sea or the Gulf of Mexico. If some catastrophe reduced humankind to a struggling remnant in a Stone Age society, or if we were wiped out entirely and some other species evolved intelligence, those resources would still be there, but any new civilization that developed would not have the tools or the material to make the machines to go out and get them, or the fuel to power those machines.

On a planet like the Earth, life may only get one shot at technology – we have exhausted the easily accessible supplies of raw materials, so if we destroy ourselves the next intelligent species, if there is one, won't have the necessary raw materials to get started. There are no second chances. And that is the last piece of evidence that completes the resolution of the Fermi paradox. They are not here, because they do not exist. The reasons why we are here form a chain so improbable that the chance of any other technological civilization existing in the Milky Way Galaxy at the present time is vanishingly small. We are alone, and we had better get used to the idea.

Further Reading

Amir Aczel, *Probablity 1*, Harcourt Brace & Co., New York, 1998
K. Akerblom, *Astronomy and Navigation in Polynesia and Micronesia*, Ethnogratiska Museet, Stockholm, 1968
Walter Alvarez, *T. Rex and the Crater of Doom*, Princeton UP, 1997
Svante Arrhenius, *Worlds in the Making*, Harper, New York, 1908
Robert Bakker, *The Dinosaur Heresies*, Longman, London, 1987
John Barrow and Frank Tipler, *The Anthropic Cosmological Principle*, Oxford UP, 1986
Steven Benner, *Life, the Universe and the Scientific Method*, Fame Press, Gainsville, 2009
Jeffrey Bennett, *Beyond UFOs*, Princeton UP, 2008
J. D. Bernal, *The Origin of Life*, Weidenfeld & Nicolson, London, 1967
Alan Boss, *The Crowded Universe*, Basic Books, New York, 2009
Ronald Bracewell, *The Galactic Club*, Freeman, San Francisco, 1974
Wallace Broecker, *How to Build a Habitable Planet*, Eldigio Press, Palisades, 1985
A. G. Cairns-Smith, *The Life Puzzle*, Oliver & Boyd, London, 1971
A. G. W. Cameron (ed.), *Interstellar Communication*, Benjamin, New York, 1963
R. M. Canup and K. Righter, *Origin of the Earth and Moon*, University of Arizona Press, 2000
Preston Cloud, *Oasis in Space*, Norton, New York, 1987
Victor Clube and Bill Napier, *The Cosmic Winter*, Blackwell, Oxford, 1990
Neil Comins, *What if the Moon Didn't Exist?*, HarperPerennial, New York, 1993
Vincent Courtillot, *Evolutionary Catastrophes*, Cambridge UP, 1999
Francis Crick, *Life Itself*, Simon & Schuster, New York, 1981
Francis Crick, *The Astonishing Hypothesis*, Simon & Schuster, New York, 1994

FURTHER READING

Ken Croswell, *The Alchemy of the Heavens*, Anchor, New York, 1995
Ken Croswell, *Planet Quest*, Free Press, New York, 1997
Michael Crowe, *The Extraterrestrial Life Debate, 1750–1900*, Dover, New York, 1999
David Darling, *Life Everywhere*, Basic Books, New York, 2001
Charles Darwin, *On the Origin of Species*, John Murray, London, 1859
Paul Davies, *The Goldilocks Enigma*, Allen Lane, London, 2006
Paul Davies, *The Eerie Silence*, Allen Lane, London, 2010
Richard Dawkins, *The Ancestor's Tale*, Weidenfeld & Nicolson, London, 2004
Jared Diamond, *The Rise and Fall of the Third Chimpanzee*, Radius, London, 1991
Jared Diamond, *Guns, Germs and Steel*, Vintage, London, 2005
Steven Dick, *The Biological Universe*, Cambridge UP, 1996
Steven Dick, *Life on Other Worlds* (abridged and updated version of *The Biological Universe*) Cambridge UP, 1998
Dougal Dixon, *After Man*, Granada, London, 1981
S. Dole, *Habitable Planets for Man*, Blaisdell, New York, 1964
Frank Drake and Dava Sobel, *Is Anyone Out There?*, Delacorte, New York, 1992
Stephen Drury, *Stepping Stones*, Oxford UP, 1999
Christian de Duve, *Vital Dust*, Basic Books, New York, 1995
Christian de Duve, *Molecules, Mind, and Meaning*, Oxford UP, 2002
Freeman Dyson, *Disturbing the Universe*, Harper & Row, New York, 1979
Freeman Dyson, *Origins of Life*, Cambridge UP, 1986
Maitland Edey and Donald Johnson, *Blueprints*, Oxford UP, 1990
Paul Ehrlich and Anne Ehrlich, *The Population Explosion*, Simon & Schuster, New York, 1990
Gerald Feinberg and Robert Shapiro, *Life Beyond Earth*, Morrow, New York, 1980
Enrico Fermi, *Atoms in the Family*, University of Chicago Press, 1954.
Ben Finney and Eric Jones, *Interstellar Migration and the Human Experience*, University of California Press, Berkeley, 1984
Richard Fortey, *Life*, HarperCollins, London, 1997
Richard Fortey, *The Earth*, HarperCollins, London, 2004
Robert Forward and Joel Davis, *Mirror Matter*, Wiley, New York, 1988
Charles Frankel, *The End of the Dinosaurs*, Cambridge UP, 1999
Louis Friedman, *Starsailing*, Wiley, New York, 1988
John Gerhart and Marc Kirshner, *Cells, Embryos and Evolution*, Blackwell, Boston, 1997

Donald Goldsmith (ed.), *The Quest for Extraterrestrial Life*, University Science Books, Mill Valley, California, 1980

Donald Goldsmith, *Worlds Unnumbered*, University Science Books, Mill Valley, California, 1997

Donald Goldsmith and Tobias Owen, *The Search for Life in the Universe*, University Science Books, Mill Valley, California, 3rd edn, 2002

Simon Goodwin and John Gribbin, *XTL*, Cassell, London, 2001

Stephen Jay Gould, *Wonderful Life*, Vintage, New York, 2000

John Gribbin, *In Search of the Multiverse*, Allen Lane, London, 2009

John Gribbin and Jeremy Cherfas, *The First Chimpanzee*, Penguin, London, 2001

John Gribbin and Mary Gribbin, *From Here to Infinity*, Royal Observatory, Greenwich/National Maritime Museum, 2008

John Gribbin and Mary Gribbin, *James Lovelock: In Search of Gaia*, Princeton UP, 2009

John Gribbin and Mary Gribbin, *Home Planet*, One World, London, 2011

David Grinspoon, *Venus Revealed*, Addison Wesley, Reading, 1997

David Grinspoon, *Lonely Planets*, HarperCollins, New York, 2003

David Harland, *The Earth in Context*, Springer-Praxis, Chichester, 2001

Nigel Henbest and Heather Couper, *The Guide to the Galaxy*, Cambridge UP, 1994

T. A. Heppenheimer, *Colonies in Space*, Stackpole Books, Harrisburg, Pa., 1977

T. A. Heppenheimer, *Toward Distant Suns*, Stackpole Books, Harrisburg, Pa., 1979

Paul Hodge, *Meteorite Craters and Impact Structures of the Earth*, Cambridge UP, 1994

Fred Hoyle, *Ice*, Hutchinson, London, 1981

Fred Hoyle and Chandra Wickramasinghe, *Evolution from Space*, Dent, London, 1981

John Imbrie and Katherine Palmer Imbrie, *Ice Ages: Solving the Mystery*, Macmillan, London, 1979

J. Jennings (ed.), *The Prehistory of Polynesia*, Harvard UP, 1979

James Kaler, *Cosmic Clouds*, Scientific American Library, New York, 1997

James Kasting, *How to Find a Habitable Planet*, Princeton UP, 2010

David Koerner and Simon LeVay, *Here Be Dragons*, Oxford UP, 2000

N. Lahav, *Biogenesis*, Oxford UP, 1999

Richard Leakey and Roger Lewin, *The Sixth Extinction*, Weidenfeld & Nicolson, London, 1996

FURTHER READING

Michael Lemonick, *Other Worlds*, Simon & Schuster, New York, 1998
Roger Lewin, *Human Evolution*, Blackwell, Oxford, 1984
James Lovelock, *Gaia: A New Look at Life on Earth*, Oxford UP, revised edn, 2000
James Lovelock, *The Ages of Gaia*, Oxford UP, revised edn, 2000
James Lovelock, *Gaia: The Practical Science of Planetary Medicine*, Gaia Books, London, 1991
Jonathan Lunine, *Earth: Evolution of a Habitable World*, Cambridge UP, 1998
Thomas McDonough, *The Search for Extraterrestrial Intelligence*, Wiley, New York, 1987
Eugene Mallove and Gregory Matloff, *The Starflight Handbook*, Wiley, New York, 1989
Lynn Margulis and Dorion Sagan, *Microcosmos*, Simon & Schuster, New York, 1986
Lynn Margulis and Dorion Sagan, *What is Life?*, Weidenfeld & Nicolson, London, 1995
Jacques Monod (tr. Austryn Wainhouse), *Chance and Necessity*, Collins, London, 1972
Simon Conway Morris, *The Crucible of Creation*, Oxford UP, 1998
Simon Conway Morris, *Life's Solution*, Cambridge UP, 2004
R. Muller, *Nemesis: The Death Star*, Weidenfeld & Nicolson, London, 1988
E. G. Nisbet, *The Young Earth*, Allen & Unwin, Boston, 1987
Gerard O'Neill, *The High Frontier*, Morrow, New York, 1977
Steve Olson, *Mapping Human History*, Houghton Mifflin, New York, 2002
Steven Pinker, *The Language Instinct*, Morrow, New York, 1994
Rudolf Raff, *The Shape of Life*, University of Chicago Press, 1996
David Raup, *Extinction*, Norton, New York, 1990
Richard Restak, *Brainscapes*, Hyperion, New York, 1995
Martin Rees, *Just Six Numbers*, Weidenfeld & Nicolson, London, 1999
Martin Rees, *Our Final Century*, Heinemann, London, 2003
Carl Sagan (ed.), *Communication with Extraterrestrial Intelligence*, MIT Press, Cambridge, Mass., 1973
Carl Sagan, *The Cosmic Connection*, Dell, New York, 1975
Carl Sagan, *The Dragons of Eden*, Random House, New York, 1977
Erwin Schrödinger, *What is Life?*, Cambridge UP, 1967 (originally published in 1944; this edition combined in one volume with the same author's *Mind and Matter*, 1958)
Robert Shapiro, *Origins*, Bantam, New York, 1987

Robert Shapiro, *Planetary Dreams*, Wiley, New York, 1999
Michael Shermer, *Why People Believe Weird Things*, Freeman, San Francisco, 1997
Iosif Shklovskii, *Five Billion Vodka Bottles to the Moon*, Norton, New York, 1991
Iosif Shklovskii and Carl Sagan, *Intelligent Life in the Universe*, Holden-Day, San Francisco, 1966
Seth Shostak, *Sharing the Universe*, Berkeley Hills Books, Albany, 1998
Seth Shostak and Alex Barnett, *Cosmic Company*, Cambridge UP, 2003
J. L. Simon, *The Ultimate Resource*, Princeton UP, 1981
George Gaylord Simpson, *The Meaning of Evolution*, Yale UP, 1967
John Maynard Smith and Eors Szathmary, *The Major Transitions in Evolution*, Freeman, Oxford UP, 1995
John Maynard Smith and Eors Szathmary, *The Origins of Life*, Oxford UP, 1999 (this is a more 'popular' version of *The Major Transitions*).
Steven Stanley, *Extinction*, Freeman, New York, 1987
Christopher Stringer and Robin McKie, *African Exodus*, Cape, London, 1996
Walter Sullivan, *We Are Not Alone*, McGraw-Hill, New York, 1966
Stuart Ross Taylor, *Destiny or Chance?*, Cambridge UP, 1998
Lewis Thomas, *The Lives of a Cell*, Viking, New York, 1974
James Trefil, *Are We Unique?*, Wiley, New York, 1997
Alan Turner and Mauricio Antón, *The Big Cats and Their Fossil Relatives*, Columbia UP, 1997
E. S. Vrba, G. H. Denton, T. C. Partridge and L. H. Burckle (eds.), *Paleoclimate and Evolution*, Yale UP, 1995
Gabrielle Walker, *Snowball Earth*, Bloomsbury, London, 2003
Peter Ward and Donald Brownlee, *Rare Earth*, Copernicus, New York, 2004
Stephen Webb, *Where Is Everybody?*, Copernicus, New York, 2002
Fred Whipple, *The Mystery of Comets*, Smithsonian Institution, Washington, DC, 1985
H. B. Whittington, *The Burgess Shale*, Yale UP, 1985
E. O. Wilson, *The Diversity of Life*, Harvard UP, 1992
Ben Zuckerman and Michael Hart (eds.), *Extraterrestrials: Where Are They?*, 2nd edn, Cambridge UP, 1995.

Index

acidification 197–8
Aczel, Amir 38
African Great Rift Valley 190
alanine 27
ALH 84001 (Mars rock) 45–6
Allen Array 41
Allen, Paul 40–41, 53
Alpha Centauri 45
aluminium-26 112
Ames experiment 29
amino acetonitrile 17
amino acids 16–17, 26–7, 30
ammonites 175
Amor (asteroid) 177–8
Andromeda Galaxy (M31) 60
angular momentum 101
Antarctica 190–91, 194
Apollo (asteroid) 177–8
Arctic, the 191–3, 197
Arrhenius, Gustaf 44
Arrhenius, Svante 44–5
asteroids
    and dinosaur extinctions 134
    and Earth collision 117
    monitoring the orbits of 200
    once part of a large object 107
    remaining planetary fragments 119
    taking energy from 98
    and the terminal Cretaceous event 172–4, 177
    threat from 198
astronomical unit (AU) 7
Aten (asteroid) 177
Atlantic Ocean 129
Australia 190

basalt 131–2
Beta Pictoris 12
Beta Taurids meteor shower 179
Big Bang, the
    definition 1
    elements produced in 3, 11
    first stars 56–7
    seeing young galaxies 55
'Big Five' (mass extinction events) 164–5
biogenesis 32–3
black holes 4, 57–8, 76
Bobylev, Vadim 203
Bracewell, Ronald 49–50, 52
Broecker, Wallace 148
Brownlee, Donald 145–6
Burgess Shale 154–7, 160, 170
butterfly effect 120

Cainozoic era 152
Caloris Basin (Mercury) 105

INDEX

Cambrian period 152–66, 172, 183
carbon
 burning in the Sun 95
 great affinity with oxygen 126–8
 unusual properties of 14–15
carbon dioxide
 effect on glaciated Earth 168–9
 greenhouse effect 97, 120, 146–7, 175
 on Mars and Venus 20–21
 regulating Earth's temperature 82–3, 131, 146–8
carbon monoxide 127
carbon-12 26
carbon-13 26
Carboniferous period 204
carnivorous birds 159–60
Ceres 107, 183
CETI 33
Chandrasekhar, Subrahmanyan 63
Chandrasekhar limit 63–4
chaos theory xiii
Chicxulub event 173–4
chimpanzees 188–90, 195
Chiron 179–80
'CHON' elements 14
chondrules 111–12
Clovis civilization 199
Clube, Victor 178–80
Cnidaria phylum see Ediacara
comet Encke 178–9
comet Shoemaker-Levy 9 122, 199
comet Wild 2 27
comets
 impacts 77, 78
 markedly elliptical orbits 103–4
 origins of 109
 and the Solar System 78
 threat to life on Earth 77–8
Communication with Extra Terrestrial Intelligence see CETI
continental drift, see plate tectonics
Continuously Habitable Zone (CHZ) 82–4
convergence 160–63
Conway Morris, Simon 151, 154, 160–63
Copernican principle 39, 41
COROT-7 9
COROT-7b 9
Cosmic Winter, The (Clube/Napier) 179
Cretaceous period 164, 172–3, 181, 185–6, 190, 199
Crick, Francis 46
Crucible of Creation, The (Conway Morris) 160

dark matter 4, 56
Darwin, Charles 28, 31, 149, 155
Darwin/TPF project 22
Darwinian evolution 158
Davies, Joel 49
Deamer, David 29
Deccan Traps, India 173, 200
deoxyribonucleic acid see DNA
Devonian period 164
dinosaurs
 cometary impact 77–8, 98, 122, 134
 death of 163–4, 176
 dominating the planet 159
 and the Palaeozoic era 73
 the Troodon 185–7, 195–6
DNA 14–17, 27–8, 31, 46, 52, 187–8, 196
Doppler effect 5, 7–8
Drake, Frank 33–6
Drake equation 33–6, 196
early bombardment 26, 28

INDEX

Earth
  birth of 66–7
  bombardment from space 134
  climate balance of 180–81
  cosmic clouds 176–80
  death of 96–7
  early catastrophe xiii
  early grip of life 26, 28, 33
  effects of a massive impact 183
  effects of the Moon 142–5
  fate of 201–3
  and the Goldilocks Zone 82
  and high obliquity 171–2
  ice ages 73, 81, 167–71, 176–7, 181, 194–5
  interglacials 194
  a living planet 100, 106
  magnetic field of 135–41
  making of 115–18
  only intelligent planet xv, 79
  origins of water 123–5
  pattern of heat from the Sun 194–5
  and plate tectonics 128–30, 132–5, 145–50
  postponing Doomsday 97–9
  presence of oxygen 20
  seeded with organic molecules 30
  stabilizing effects of the Moon 142–5, 149, 171–2
  temperature regulated by carbon dioxide 82–3, 131
  threat to life from comets 77–8
  unique disposition of water 133
  very little carbon 127
  and volcanism 200
  vulnerability to sudden shocks 197
Earth System Science *see* Gaia
East African Rift Valley 192–3
Ediacara 153, 155, 161
Encke, Johann 178
Encke's Comet 178–9
enzymes 27–8
equilibrium 28
Eros (asteroid) 177
ethyl formate 14, 17
eukaryotic cells 152–3, 167–8, 170
exoplanets *see* extrasolar planets
Extinction (David Raup) 202
extrasolar planets
  around pulsars 5–6
  around Sun-like stars 6–7, 10–11
  description 4–8
  detection of 20, 22–4
  formation of 13
  search for evidence of life on 22
  where carbon dominates over oxygen 127
  see also hot jupiters

faint young Sun paradox 21
Fermi, Enrico 41–3, 48, 50, 53
Fermi paradox 41–3, 46–7, 68, 79, 187, 205
'Fermi questions' 42
51 Pegasi 6–7
5–carbon sugar, *see* ribose
ribose 16
Forward, Robert 49
fossil fuels 204
fossils 27
Frail, Dale 6

Gaia xiv–xv, 19–22, 24–5, 83, 100, 197
galactic club 47–8
Galactic Habitable Zone (GHZ) 74–9, 85, 89
galaxies

formation of early 58–60
formation of spiral patterns 71
other 2
Galaxy, the *see* Milky Way Galaxy
gamma rays 76
Gemini Observatory, Hawaii 10
genetic code 15
GHZ, *see* Galactic Habitable Zone
Gliese 581 10
Gliese 710 203
Glikson, Andrew 134
global thermostat 146
global warming 197
global winter 175
glycine 17, 27
glycolaldehyde 15–16
Goldilocks Zone xiv, 78, 82
gorillas 189–90, 195
Gott, Richard 38–40
Gould, Stephen Jay 151, 154, 156–60, 166
graphite 127
gravity 62, 64, 69
greenhouse effect 97, 120, 146–7, 168, 175

half-life timescale 90
Hallucigenia 155, 160
Hart, Michael 43, 50
HD 189733 10, 18
helioseismology 92
helium 12–13, 94–5, 108
Hephaistos 179
Herschel infrared telescope 116
Hertzprung-Russell Diagram 84, 86, 95
Hipparcos satellite 203
HL Tauri 12
Homo erectus 40, 184, 193–4

Homo sapiens 40, 187, 192
hot jupiters xiv, 5–10, 18, 101, 103, 110
 see also extrasolar planets
How to Build a Habitable Planet (Broecker) 148
Hoyle, Fred 30, 180–81
HR 4796A 18–19
Hubble Space Telescope 18, 58
hydrogen 12–13, 15, 65, 94–5, 108, 123, 177
hydroxyl radical 123

Ice Ages 73, 81, 167–71, 176–7, 181, 194–5
Iceland 129
ichthyosaurs 162, 164
infrared radiation 8, 10–12, 20, 22–4, 56, 68
inspection paradox 38–9
Integral space telescope 68
intelligence
 and convergence 162–3
 and Drake 35
 emergence of 2, 25
 evolvement on Earth 61, 106, 122, 125, 132
 hard to define 185
 interference of comets 77, 77–8
 overwhelmingly improbable event 158
 and primates 48–9
 rarity of xiv–xv, 79, 82, 84–5, 94
 and Von Neumann 51
interferometry 23
inverted cone of life 156–7, 160–61
iridium layer 173–5
iron 62–4, 65
iron-60 112
Jones, Eric 42

INDEX

Jupiter
   angular momentum of 101
   beneficial effects of 119–24, 127
   and the close encounter 103
   description/key to life 107–10, 114–15
   effect on Earth's orbit 194
   size of 7
   see also hot jupiters
Jurassic period 162

kangaroos 162
Kasting, James 183
KBOs *see* Kuiper Belt
Kepler space observatories 8
Konacki, Maciej 6
krypton 124
Kuiper Belt 98–9, 108–10, 115, 122, 178

Lagrangian points 116–17
Lake Toba, Indonesia 200
Large Magellanic Cloud 91
Late Heavy Bombardment 32–3, 122, 124, 127, 133, 140
Leakey, Richard 165
Lepock, James 45
Life Itself (Crick) 46
Life's Solution (Conway Morris) 162–3
Lin, Douglas 113
Lineweaver, Charles 31–2, 35, 74, 77
lipids 29
Local Arm 70, 73
look back time 55
Lovelock, James xiv, 19–22, 24, 83, 97, 100, 148, 197

M13 cluster 33–4

magnetosphere *see* Earth, magnetic field of
Main Sequence *see* Hertzprung-Russell Diagram
mammals 157, 159
Margulis, Lynn 28, 153
Mars
   lack of biology 100, 106
   lack of Earth-like geological features 139–41
   and Lovelock 20–21
   origins 115
   size problems 119–20
   and the Sun 96
   temperature of 82
marsupial moles 162
Maynard Smith, John 163
Mayor, Michel 7
Mediterranean Sea 191
Meert, Joseph 170
Melosh, Jay 45, 134, 175
Mercury 7, 85, 96, 104–5, 115, 118, 121
Mesozoic era 152
meteorites
   and amino acids 17, 26, 28, 44
   cosmic rubble 107
   and the death of the dinosaurs 164
   from Mars 45–6
   impacting Venus 140
   presence of calcium/aluminium 93
   presence of chondrules 111–12
   presence of nickel-60 91
methane 18–19
Milankovitch, Milutin 194
Milankovitch Model 194–5, 197
Milky Way Galaxy
   black hole at the centre 58, 68, 75–6

INDEX

carbon-to-oxygen ratios 126, 128
and cosmic cannibalism 59
four categories of stars 67–9
future colonization of 50, 52–3
Galactic Habitable Zone
  (GHZ) 74–9, 85, 89
and gamma ray bursts 76
giant molecular clouds in 86–8
as a giant warehouse 19
habitable planets xiii, xv, 34–6
prospect of other technological
  civilizations within 205
size of 2, 11
spiral structure 3–4
star formation disturbances 71–2
stars with dusty discs 102–3
stars other than the Sun 84
the Sun's place 69–70, 73–4, 104
why intelligent life exists within
  55
Miocene age 190
Mirror Matter (Forward/Davies) 49
molecular clouds 17–18
molluscs 162
Monod, Jacques 184
Moon, the
  benefits of being large 141–2
  effects on Earth 142–5
  formation of xiii, 32, 116–18
  impact evidence used as a
    measure 134
  life-giving role 139–40
  stabilizing effects on the Earth
    142–5, 149, 171–2
  uniqueness of 150
Morgan, Joanna 174–5
Morris, Conway 154–6, 166
Murchison meteorite 29
n-propyl cyanide 13, 17
Napier, Bill 178–80

NASA 20–22, 29–30
Nebulae
  Orion 11
  Rosette 102–3
Nectocaris 154–5
negative feedback 147
Neptune 1–2, 103, 108–10, 115
neutrons 63
nitrogen 131, 146
Nix Olympica (Mars) 141

oceanic/continental crust 130
Oort, Jan 77
Oort Cloud 77–8, 109, 122, 178,
  203
Opabinia 155
Ordovician period 164
Orgel, Leslie 46
Orion Arm *see* Local Arm; Orion
  Nebula
Orion Nebula 11
oxygen 13, 15, 20, 24, 123, 126–7
ozone 22, 24, 123, 175, 177

Pacific Ocean 129
Palaeozoic era 73, 152
Pallas 107, 183
Pangaea 165
panspermia 33, 43, 45
  directed 33, 46
Pasteur, Louis 44
pentadactyl limb 161
Permian extinction event 173
phyla/phylum 157–8, 160, 162, 164
planetisimals 12
planets
  making of 109–13
  orbits of 103–4
  in other systems 102
  rocky 100–101

see also hot jupiters; extrasolar planets
plate tectonics 128–30, 132–5, 145–50
Pliocene age 190
Polynesia 49
Pompeii worms 81
PPDs *see* proto-planetary discs
Precambrian period 152, 155, 158, 166, 172
Principle of Terrestrial Mediocrity *see* Copernican principle
probability theory 37
prokaryotes 152–3, 155, 167
protein 16
proto-planetary discs (PPDs) 11–13, 102
protons 63
Proxima Centauri 103
pulsars 5–6

Quaternary Period 162
Queloz, Didier 7

R136 cluster 91
radioactive isotopes
   aluminium-26 90–91
   iron-60 90–91, 93
   nickel-60 90–91
Raup, David 202–3
Rees, Sir Martin 37, 48
retrograde orbits 101
ribonucleic acid *see* RNA molecules
Ribose 15
RNA molecules 15–16, 27–8
Rosette Nebula 102–3
Russell, Dale 186

sabre-toothed big cats 162

Sagan, Carl 186–7
Saturn 108–10, 115
Search for Extra Terrestrial Intelligence *see* SETI
Secker, Jeff 45
Seguin, Ron 186
semi-permeable membranes 28
serine 27
SETI 34, 40–41, 128–9, 187
Shermer, Michael 35–6, 38
Shoemaker-Levy 9 comet 122, 199
Siberian Traps 173
silicon 126, 128
silicon carbide 18, 127
silicon dioxide (silica) 131
snowline 113
Sodium 65
Solar System
   an orderly place 98–9
   carbon mostly blown away 127
   circularity of orbits of planets 103
   close supernova explosion 91
   and comets 78
   and cosmic clouds 176–80
   effects of gas cloud passage 72–3
   formation of 2–3, 66, 70, 92, 101, 113–15
   formation of the rocky planets 111
   geography/planets of the 7, 104–9
   and the Goldilocks Zone 82
   a special place 104, 118–25
solar twins 93–4
South America 190
spiral arms 3, 69–70, 72–4, 78
spiral density waves 71–2

INDEX

Spitzer Space Telescope 10, 18, 101–3, 150
squirrel monkeys 189
Stanley, Steven 165
star types
  A 86
  D (white dwarfs) 62–4, 96–7
  F 86
  G (yellow-orange) 8, 84
  K 86
  M (red dwarfs) 10, 84–5
  O and B 86, 102–4
  red giants 95–7
star-forming regions 11
Stardust spaceprobe 27
stars
  and angular momentum 87–8
  with companions 86, 88–9
  formation disturbances 71
  main sequence 61
  metallicity of 65–6, 68, 74–5, 79
  in the Milky Way 67–70
  neutron 5–6, 64
  and nucleosynthesis 126
  Population I 59, 67
  Population II 57, 67
  pre-stellar cores 87–8
  supernovas *see* supernovas
Stellar Habitable Zone (SHZ) 80–82
stellar nurseries 87
Stevenson, David 149
stromatolites 27
Sturtian glaciation 170, 172
sugars 26, 30
sulphuric acid 175
Sun, the
  close supernova explosion 91

Continuously Habitable Zone (CHZ) 82–4
death of 94–7
encounter with a star 104
fate of 63, 201–2
gravitational pull of 116
metallicity of 66, 92–4
not an average star 84–6, 90
ordinary star 2
place in the Milky Way 69–70, 79
radioactive elements of 90–91
rotation of 101
solar eclipses 143
and solar storms 137–8
Stellar Habitable Zone (SHZ) 80–82
and stellar nucleosynthesis 3
structure of 60–62
tides on Earth 144–5
Water vapour from energy of 180–81
yellow-orange G-type star 8, 84
supernovas
  description 62–5
  explosion close to the Sun 91, 93, 112
  and giant molecular clouds 87
  in the Milky Way Galaxy 75
  pulsars 5–6
  in spiral arms 72

Taurids meteor shower 179
terminal Cretaceous event 164–5, 172–4, 177, 185–6, 198
Theia (proto-planet) 116–17
tholins (organic compounds) 19
Thomson, William 43–5
Tipler, Frank 52
Titan 19

Tommotian 153
'tree of life' 157
Triassic period 164, 181
Trigo-Rodriguez, Josep 112
Tunguska event 198–9
Turing, Alan 51
23 Lib 119
2–carbon sugar *see* Glycolaldehyde
Type I supernovas 63–4
Type II supernovas 64–5
Tyrannosaurus Rex 159

ultraviolet radiation 123, 175
Unidentified Flying Objects (UFOs) 470
universal Turing machine 51
Upsilon Andromedae 121
Uranus 108–10, 115, 182–3

Venus
    carbon dioxide atmosphere 21
    catastrophic event xiii
    cratering record puzzle 181–2
    early formation 115
    effects of gravitational interactions 121
    fate of 96
    greenhouse effect 120, 146
    lack of Earth-like geological features 139–41
    no life 100, 105
    reverse rotation of 182–3
    too hot for liquid water 82
vesicles 29–31
Vesta 107, 183
Viewing, David 43
von Neumann, John 51–2

von Neumann machines 52–3, 98

Ward, Peter 145–6
water
    in discs around stars 12
    in Earth's oceans 148
    essential for plate tectonics 135
    in hot jupiters 10
    origins of on Earth 122–5
    and plate tectonics 131–2, 145–6
    unique disposition on Earth 133
    vapour from Solar energy 180–81
    vital for life 80–81
Webb, Stephen 47
Wesson, Paul 45
wet earths 11
Where is Everybody? (Webb) 47
Whipple, Fred 179
Whittington, Harry 155
Wickramasinghe, Chandra 30
Wild 2 comet 27
WISE satellite 200
Wolszczan, Alex 6
Wonderful Life (Gould) 154, 156, 160

X-rays 68
xenon 124

Yellowstone Park, United States 200
Yucatan Peninsula 173–4

zodiacal light 180
Zwart, Simon 92–3